'The perfect bedside book for anyone who wan  of physics' Robin Ince

'*The Matter of Everything* is a magical tour of incredible century in physics, which saw explor tabletop curiosities to literally the world's larges

'The stories in and of themselves are fascinating, science: the power of international collaboration and competition, the generations it takes for science to translate from the lab to social benefit, the interdependency between basic and applied, and the unwritten contribution of many women scientists' Jonathan Grant

'*The Matter of Everything* is an impassioned, elegant history of particle physics and its applications. Dr Sheehy adroitly brings together a glittering cast of characters – from the famous giants of the field to the unjustly overlooked scientists whose shoulders they stood on' Ananyo Bhattacharya

'Sheehy's attention to detail shines through every story and yet there's a lightness of touch in the way she highlights the passion, drive, ingenuity and, ultimately, the sheer triumph of science in unlocking nature's secrets' Jim Al-Khalili

'A magnificent tour of the experiments that have shaped both how we see our world but also the way it functions. It places you right at the critical moment in these labs alongside scientists famous and unknown, effortlessly explaining their amazing discoveries and the world-transforming impacts' Alan Duffy

'This is a book about the fundamental problems of physics written from a viewpoint I hadn't come across before: that of the experimenter ... The actual sequence of experiments, and failures, and more experiments, and success, is a fascinating one, especially for any readers excited, as I am, by the thought of making things' Philip Pullman

'*The Matter of Everything* is a wonderful telling of the history of particle physics ... The book is so much easier to read than many other tomes on this subject – Suzie has written a book that makes these important stories and ideas from the history of science accessible, and genuinely enjoyable to read' Alom Shaha

'Science has made many important discoveries throughout history. But sometimes how these discoveries came about is even more interesting than the discoveries themselves, a point *The Matter of Everything* makes brilliantly. A fascinating book' Dean Burnett

'It's a rare writer that can pluck the biggest ideas in science out of the sky, and bring them down to earth in a way that anyone can understand ... Suzie takes on the seemingly unreachable inner life of atoms, and places it directly into our hands. More than just a history of particle physics, this is an interconnected web of experiments, people and stories that are simultaneously passionate and profound' Helen Arney

'A lovely book. Well written. Reflective. Informative. Balanced. Great stuff. I wish other physicists would be as clear' Dr Markus Nordberg

'A celebration not only of science but also of scientific collaboration ... It's delightful to be a fly on the wall as great names such as Ernest Rutherford and J.J. Thomson – and many less well-known but no less impactful – pace their laboratories, hunch over equipment often cobbled together from literal string and sealing wax, argue loudly with each other in print and at conferences and scratch their heads as an experiment yet again delivers an unexpected result' *Spectrum*

'This fascinating and pacey book introduces a century of investigators who tested theory against reality. Through a combination of persistence and luck they discovered x-rays, split the atom and transformed the world and our understanding of it' *Sunday Times*

DR SUZIE SHEEHY is a physicist, science communicator and academic who divides her time between her research groups at the University of Oxford and the University of Melbourne. Her research addresses both curiosity-driven and applied areas, and is currently focused on developing new particle accelerators for applications in medicine. *The Matter of Everything* is her first book.

suzie sheehy

the matter of everything

A
History
of
Discovery

BLOOMSBURY PUBLISHING

LONDON · OXFORD · NEW YORK · NEW DELHI · SYDNEY

BLOOMSBURY PUBLISHING
Bloomsbury Publishing Plc
50 Bedford Square, London, WC1B 3DP, UK
29 Earlsfort Terrace, Dublin 2, Ireland

BLOOMSBURY, BLOOMSBURY PUBLISHING and the Diana logo are trademarks
of Bloomsbury Publishing Plc

First published in Great Britain 2022
This edition published 2023

A catalogue record for this book is available from the British Library

ISBN: HB: 978-1526-61896-2; TPB: 978-1-5266-1895-5; PB: 978-1-5266-1899-3;
EBOOK: 978-1-5266-1898-6; EPDF: 978-1-5266-4710-8

2 4 6 8 10 9 7 5 3 1

Typeset by Newgen KnowledgeWorks Pvt. Ltd., Chennai, India
Printed and bound in Great Britain by CPI Group (UK) Ltd, Croydon CR0 4YY

MIX
Paper | Supporting
responsible forestry
FSC® C171272

To find out more about our authors and books visit www.bloomsbury.com
and sign up for our newsletters

CONTENTS

Introduction   1

PART 1   DISMANTLING CLASSICAL PHYSICS
    1   Cathode Ray Tube: X-rays and the Electron   11
    2   The Gold Foil Experiment: The Structure of the Atom   29
    3   The Photoelectric Effect: The Light Quantum   45

PART 2   MATTER BEYOND ATOMS
    4   Cloud Chambers: Cosmic Rays and a Shower of New
          Particles   71
    5   The First Particle Accelerators: Splitting the Atom   95
    6   Cyclotron: Artificial Production of Radioactivity   119
    7   Synchrotron Radiation: An Unexpected Light Emerges   139

PART 3   THE STANDARD MODEL AND BEYOND
    8   Particle Physics Goes Large: The Strange Resonances   159
    9   Mega-detectors: Finding the Elusive Neutrino   181
   10   Linear Accelerators: The Discovery of Quarks   199
   11   The Tevatron: A Third Generation of Matter   219
   12   The Large Hadron Collider: The Higgs Boson and
          Beyond   243
   13   Future Experiments   265

Acknowledgements   277
Notes   279
Index   303

Introduction

Some years ago, I was sitting at a laptop frowning at the seemingly easy question I'd just been asked by four particle physics professors from the University of Oxford. I had missed their names, not just from nerves but because my PhD interview was conducted over an unstable internet connection from a motel room in outback Australia. They had asked me: 'what do you find fascinating about particle physics?'

This was a trick, surely: Oxford admissions interviews are notoriously tough. In that moment, I decided it was best to be honest. I told them of my wonder at the way physics seemed to be able to describe everything: from the smallest subatomic particles to the atoms that make up our bodies, up to the largest scales of the Universe, and how all of this was connected.

Particle physics, I said, was the foundation of it all.

Five years earlier, I had been studying civil engineering at Melbourne University. I had never known that being a physicist was an option: while I'd enjoyed physics in school, I'd only ever known of it leading to a career in engineering. That all changed a year into my university degree, when I was invited along by my classmates to the annual highlight of the physics student society calendar: astro camp.

One Friday afternoon we left Melbourne and arrived, two hours later, at the Leon Mow Dark Sky Site. The bumpy dirt road led us to a tin-roofed building where we unpacked beer and telescopes, then set up our tents near a large clearing. As the light faded, the temperature dropped and the sound of cicadas began to pierce the air. To preserve my night vision, I used a hair tie to hold a piece of red cellophane over my torch light. I clambered into my sleeping bag, grateful for its dual function as a source of warmth and an insect barrier. I breathed in the familiar scent of gum trees. Then I looked up.

'There's one!' the man next to me shouted, as a meteor blazed across the sky. As my eyes adjusted to the darkness, the true wonder of this designated 'dark sky site' revealed itself. The chatter fell to whispers, which in turn fell to a hush. Venus slowly set below the horizon and other planets came into view. Over the course of that night, I got a sense of the slow but constantly changing nature of the night sky. Through my friends' telescopes I saw the magnificent rings around Saturn, familiar from pictures but strangely new through a lens, stars forming in nebulae full of glowing dust and globular clusters sparkling with millions of stars orbiting our galaxy 100,000 light years away.

The most spectacular view was the bright band of stars and dust, the glowing arc of our own galaxy, the Milky Way. From the Southern Hemisphere, we look towards the middle of our disc-shaped galaxy. We're about two-thirds of the way out from the middle, orbiting our star, which itself is moving within the Milky Way. The galaxy is cruising through space along with its local group of galaxies at about 600 kilometres per second. Beyond it are billions more like it, stars and nebulae, black holes and quasars, matter formed from energy transformed through immense tracts of space and time.

That moment was when I truly grasped how small I was, how short-lived, and how I struggled to put words to the magnitude of what I was seeing. The stars and planets weren't *up there* and I wasn't *down here*: it was all part of one enormous physical system called the Universe. I was a part of it, too. Of course I knew that already, but I'd never really *felt* my place in it until that moment.

Suddenly, nothing else mattered. I wanted to know more, about gravity and particles and dark matter and relativity. About stars and atoms and light and energy. Above all, I wanted to know how it was all connected and how I was connected with it. I wanted to know if there really was a theory of everything. I felt deeply that all this mattered, that it mattered to me as a human, that understanding this was a goal big enough that if I managed it even a little bit, I'd not have wasted my blip of time as a conscious being. I decided to become a physicist.

The aim of physics is to understand how the Universe and everything in it behaves. One of the ways we try to do this is to ask questions, and as I studied more physics the question that seemed to lie at the core of it all was: 'What is matter, and how does it interact to create everything around us – including ourselves?' I suppose I was trying to figure out the meaning of my own existence. Rather than study philosophy, I went about it in a more indirect way: I set about trying to understand the entire Universe.

People have asked questions about the nature of matter for millennia, but only in the last 120 years has this curiosity finally led us to some answers. Today our understanding of the tiniest constituents in nature and the forces that govern them is described by the field of particle physics, one of the most awe-inspiring, intricate and creative adventures that humans have ever embarked on. Today we have intimate knowledge of the physical matter of the Universe and how it all fits together. We've found that reality has a richness and complexity that humans just a few generations ago could never have imagined. We've overthrown the idea of atoms as the smallest bits of our world and discovered fundamental particles that play no role in ordinary matter, but appear necessary based on the mathematics that – somewhat miraculously – describes our reality. In just a few decades we have learned how to fit all of those pieces together, from the blast of energy at the start of the Universe to the most precise measurements in nature.

Our view of the smallest constituents in nature has changed rapidly throughout the last 120 years: from radioactivity and the electron, to the atomic nucleus and the field of nuclear physics, together with the development of quantum mechanics (which describes nature on the smallest scales). Some way into the twentieth century this work became known as 'high energy physics', as new particles were found and the focus shifted away from the atomic nucleus. Today the study of all the many particles and how they are formed, behave and transform is simply called particle physics.

The Standard Model of particle physics classifies all the known particles in nature and the forces through which they interact. It was developed by many different physicists over decades and our current version came about in the 1970s. This theory is an absolute triumph: it is mathematically elegant and unbelievably precise, yet fits on the side of a mug. As a student, I was drawn in by how completely the Standard Model seemed to describe how nature works at a fundamental level.

The Standard Model tells us that all the matter that makes up our everyday existence is composed of just three particles. We consist of two types of quark called 'up' and 'down' which form our protons and neutrons. These two quark types together with electrons make up atoms, held together by forces: electromagnetism and the strong and weak nuclear forces. That's it. That's us and everything around us.[1] Yet despite being composed of nothing more than quarks and electrons, we – humans – have somehow figured out that there is *much more* to nature than this.

Our triumph of knowledge has not come about purely through conceptual and theoretical leaps. The stereotype of a lone genius theorising at a desk is largely incorrect. For over a century, questions like 'What is inside the atom?', 'What is the nature of light?' and 'How did our Universe evolve?' have been addressed by physicists in an entirely practical way. The reason we can say today that we *know* all this stuff, that we think our theoretical models represent reality, is not because we have pretty mathematics but because we have done experiments.

While many of us come across the idea that protons, neutrons and electrons make up the world around us as children, very little is said about *how* it is that we learned about matter and forces and, well, everything. A proton is a million million times smaller than a grain of sand and it is far from obvious how we actually go about working with matter on so small a scale. This is the art of experimental physics: to follow our curiosity from a seed of an idea, to a real physical piece of equipment, to the accumulation of new knowledge. That evening at the dark sky site, that understanding that I enjoyed physics more when I got to experience it firsthand, led me towards the idea of being an experimental physicist.

While theoretical physicists can revel in mathematical possibility, experiments take us to that frightening frontier of vulnerability: the real world. This is the difference between theory and experiment. While a theoretical physicist's ideas must take into account the results of experiments, an experimental physicist has a more nuanced job. She is not simply testing out the ideas of theoretical physicists; she is asking her own questions and designing and physically building equipment which she can use to test those ideas.

The experimentalist must understand and be able to use theory, but she must not be constrained by it. She must stay open to finding something unexpected and unknown. She also has to understand many other things: her practical knowledge ranges from electronics to chemistry, from welding to handling liquid nitrogen. She must then combine these things together to allow her to manipulate matter that she cannot see. The truth is that experiments are hard, and the process involves many false starts and failures. It takes a certain kind of curiosity and personality to want to do this. Yet throughout history, many have had the passion and persistence to do so.

Over the last century the experiments that scientists have used in particle physics have gone from single-room setups led by one person to the largest machines on Earth. The era of 'Big Science', which began in the 1950s, has now grown to produce experiments that involve collaborations of over a hundred countries and tens of thousands of scientists. We build underground particle colliders consisting of many kilometres of high-precision electromagnetic equipment in projects that span more than twenty-five years and cost billions of dollars. We have reached a point where no individual country can achieve these feats alone.

At the same time, our everyday lives have gone through a similarly dramatic transformation. In 1900, most homes were twenty years away from having electricity, horses were the main form of transport and the average lifespan in the UK or United States was less than fifty years. Today we are living longer, in part because when we get sick the hospital has MRI, CT and PET scanners to help diagnose illnesses and a range of medicines, vaccines and high-tech gadgets to treat us. We have computers, the World Wide Web and smartphones to connect us, which have created entirely new industries and ways of working. Even the goods around us are designed, augmented and enhanced using new technology, from the tyres on our cars to the gemstones in our jewellery.

When we think about the ideas and technologies that make up the modern world, we rarely associate this with the parallel trajectory of experimental physics, but they are intimately connected. All the examples above came about from experiments designed to learn more about matter and the forces of nature – and this list is only scratching the surface. Within just two generations, we have learned to control individual atoms to build computing devices so small that even a microscope struggles to see them; to use the unstable nature of matter to diagnose and treat disease; and to see inside ancient pyramids using high-energy particles from space. It's all possible because of our ability to manipulate matter at the level of atoms and particles, knowledge which came from curiosity-driven research.

I chose to be an experimental physicist in the field of accelerator physics: I specialise in inventing real-world equipment that manipulates matter on this tiny scale. Accelerator physicists constantly discover new ways of creating beams to help learn more about particle physics, but increasingly our work contributes to other parts of society. It still surprises students, friends and audiences when I tell them that their nearest hospital almost certainly houses a particle accelerator, that their smartphone relies on quantum mechanics and that their ability to browse the web is only possible because of particle physicists. We build particle accelerators to study viruses, chocolate and ancient scrolls. Our detailed understanding of the geology and ancient history of our planet is the outcome of research in particle physics.

Curiosity-driven research takes us past the limits of what we know and what we expect, leading us to ideas, frontiers and solutions that change the course of history. Through this search for new knowledge, we bridge the gap between what we know to be possible and what we believe to be impossible. That is where curiosity leads to truly groundbreaking innovation. Physics, in particular particle physics, offers perhaps the most striking examples of this phenomenon. So how did a series of physics experiments lead to all of these aspects of our modern world?

There have of course been thousands of experiments, all of which have contributed to our knowledge in some way. In this book, I will take you through twelve key experiments that marked a first – a discovery – that we now see as essential to our understanding of the world we live in. We will begin with experiments conducted by a few individuals in small labs in England and Germany at the turn of the nineteenth century – experiments that showed classical physics breaking down, alerting us to the existence of entities smaller than atoms. From there we will see how experiments in Chicago helped validate the emerging ideas of quantum mechanics, leading physicists around the world to soar in hot-air balloons and hike up mountaintops on the trail of new particles. Each experiment reminds me of the co-mingling of frustration and joy that I know all too well from my own lab, that very human experience of doing hands-on science, but the benefit of hindsight allows me to see what these early experimenters could not: what became of their discoveries and inventions.

The next experiments take us to the race between the United States, Germany and the UK to build the first particle accelerator and split the atom. These experiments were involved in the creation of artificial radioactive elements in California, and led to a serendipitous discovery by industrial scientists that created both a new tool for research and a new understanding of astronomy. Finally, we follow the stories of teams and nations joining together to build the big experiments that formed the background of my own career: from US labs like Brookhaven and Berkeley to the Stanford Linear Collider and Fermilab and, eventually, to the European Organisation for Nuclear Research (CERN).

Taken together, these experiments embody the spirit of enquiry that stems from human curiosity. Over the course of a century they have changed our lives in almost every aspect, from computing to medicine, from energy to communications and from art to archaeology. Physics will always be, at its core, about understanding our place in the Universe, a truth I have felt ever since I saw the night sky anew. This journey illustrates how physics has also led to so much of the modern technology we now take for granted, and to practical outcomes we never even imagined. It tells us that doing physics has something to teach us all about curiosity and the power we all have to make breakthroughs that might change the world.

Part 1

Dismantling Classical Physics

Imagination is the discovering faculty pre-eminently. It is that which penetrates into the unseen worlds around us, the worlds of science. It is that which feels and discovers what is, the real which we see not, which exists not for our senses.

– Ada Lovelace, in a letter to
Lord Byron, January 1841

1

Cathode Ray Tube: X-rays and the Electron

Our story begins in a laboratory in Würzburg, Germany in 1895. It didn't look much like the clean white spaces used by modern scientists; it had beautiful parquet floors and impressive high windows looking out over the park and vineyards opposite. The physicist Wilhelm Röntgen closed the shutters and turned to his work. On a long wooden table, he set up a glass tube the size of a small wine bottle, which had most of the air removed using a vacuum pump.[1] Wires trailed off from metal electrodes, one in the end of the tube (the negative cathode) and one roughly halfway down the length (the positive anode). When high-voltage electricity was applied, a glow appeared inside – the so-called 'cathode rays' that gave the tube its name. So far, everything was as he expected. Then, out of the corner of his eye, he noticed a small screen on the other side of his lab glowing.

He walked over to inspect it. The phosphor-coated screen was giving off a green-coloured light. When he turned the cathode ray tube off, the light disappeared. When he turned the tube back on, the light returned. Perhaps it was just a trick of the eye, a reflection of the light from the glowing cathode ray tube? He covered the tube with some black cardboard but found that the light on the screen persisted. He'd never seen anything like it before, but thought it could be important.

From this moment on, physics would never be the same. Beginning with this first serendipitous observation, experiments using cathode ray tubes would lead the field of physics into entirely new territory and start to overturn ideas about the natural world that had been accepted for millennia. In time, the cathode ray tube would lead to technologies which changed the way people live, work and communicate. It all started here, with this glowing screen, and the curiosity of an individual.

Wilhelm Röntgen, like most scientists around the world at the end of the nineteenth century, agreed that the subject of physics was almost complete. The Universe was made of matter that consisted of 'atoms'. They'd figured out that there were different types of atoms, which corresponded to different chemical elements. From trees to metals, water to fur: all the complexity of the material world around them differed in terms of hardness, colour and texture because they were built of different atoms, which they viewed like tiny, spherical Lego pieces. If you had the right instructions, you could take a particular set of atoms and create anything you liked.

They also knew there were forces through which everything interacted. Gravity kept the stars in our galaxy, and our planet circling the Sun. Even the mysterious forces of electricity and magnetism had finally been brought together into just one force: electromagnetism. The Universe was predictable: if you had all the details of the inner workings and set things in motion, the movements of all matter could be predicted perfectly.

Now only the details were left to explore – details like how exactly the cathode ray tube worked, one of the few small things they couldn't quite explain. There were theories of course, including the idea that the glow inside was related to ripples in the hypothetical aether, the medium through which light was thought to travel in much the same way as sound is transmitted by the air. Now, in his investigations of the details of the cathode ray tube, Röntgen seemed to have stumbled onto a complication. Not only was there something unexplained happening inside the tube, but he'd found a strange effect happening on the outside as well.

Röntgen had seemed ordinary as a child. The son of a cloth merchant, he loved exploring nature in the countryside and forests.[2] The one thing he did show quite an aptitude for was making mechanical things[3] and this early ability turned out to be useful to his experimental work later in life. As an adult, his dark hair stood up from his forehead 'as if he were permanently electrified by his own enthusiasm'.[4]

Röntgen was a shy man who gave lectures in an intolerably low voice, was strict with his students and was even slightly uncomfortable at the idea of having assistants in his lab. But he loved science, sometimes quoting the great engineer Werner von Siemens, who said 'The intellectual life gives us at times perhaps the purest and highest joy of which the human being is capable.'

Now he had found something that no one had seen before. When he saw the strange glowing screen, he assumed that he wasn't looking at the same kind of 'ray' which caused the cathode ray tube to glow, since that effect seemed contained inside the tube. Instead he'd found a new kind of invisible ray which seemed to be able to travel much further. He immediately dedicated himself to exploring more, channelling all his time and energy into the lab. When later asked what he thought at the time he said 'I didn't think, I investigated'. He had a number of similar tubes around his laboratory[5] which he could now use with the phosphor screen, setting up each in a methodical and thorough way to figure out the nature of the new rays. He placed different materials between the tube and the screen, trying paper, wood and even hard rubber. The rays went through all of them, barely diminishing. When he pointed the rays through the thick wooden door to the adjoining lab, he found he could detect them on the other side. Only when he placed aluminium foil in front of the tube did the rays seem to have some difficulty getting through.

He spent seven intense weeks in his lab, occasionally being reminded to eat by his wife, Anna Bertha. Apart from those interactions, he was working almost entirely alone, and he remained silent about his research. He didn't tell his assistants, let alone his international colleagues. He knew that if he didn't announce his discovery first, hundreds of other scientists who had similar experiments sitting in their labs would beat him to it. The only report of him speaking about the work was to a good friend, to whom he simply said 'I have discovered something interesting but I do not know whether or not my observations are correct'.[6]

Next, he tried sticking his hand in the way of the rays and reported: 'If the hand is held between the discharge tube and the screen, the darker shadow of the bones is seen within the slightly dark shadow-image of the hand itself...' This gave him an idea. He used the rays to make an image of Bertha's hand on a photographic plate, which confirmed his understanding: the rays travelled easily through the skin and flesh but not so easily through bone or metal. The bones in her hand and her wedding ring showed up dark in contrast to the flesh that we normally see with the eye. The ability to block the new rays was related to the density of an object. According to legend, when Bertha saw the bones in her hand she exclaimed 'I have seen my death!' and never set foot in her husband's lab again.

Röntgen needed to give the new rays a name in his notebook. In science, we typically denote things which are unknown with a letter like 'X', and so Röntgen came up with possibly the best unintentional branding in the history of physics. He called his new discovery 'X-rays'.

Once he was satisfied that he understood how X-rays behaved, Röntgen had a decision to make. Should he patent the idea, publish his findings, or do more work before he announced his discovery? There were many questions that he was still curious about, like how X-rays were related to light and matter, what they were made from and how they were formed. He determined that he couldn't delay the announcement any longer; the chance of someone else finding X-rays was too high. If he published the discovery before applying for a patent, he would never make any money from it if it turned out to be useful in medicine. But Röntgen was a physicist, not a doctor, so he didn't know if medics would be interested in his idea or not. He decided the best way to make it useful was to publish his discovery and communicate it to the medical community.

Overcoming his habitual shyness, on 23 January 1896, Röntgen set up a heavy table with his X-ray experiment in the Würzburg Physical Medical Society lecture theatre, just a short walk from his laboratory. The crowd had already caught wind of his discovery through newspaper articles and so many attended that there were men standing in the aisles. Röntgen presented the first ever lecture about what he'd discovered. He showed the audience how X-rays could go through wood and rubber, but not through metal. He showed them the photograph of Bertha's hand and told them about his idea to use X-ray pictures to see inside the human body. To drive the point home, he decided he would demonstrate just how easy it was to create a similar image.

Standing in front of the hall, he invited the president of the society, a prominent anatomist, to place his hand in the path of the X-rays. Röntgen switched on the cathode ray tube and took an X-ray photograph of the president's hand. The doctors in attendance were amazed. They immediately saw the value of his discovery and the president was so impressed that he led the crowd in giving Röntgen three cheers. They even proposed to name the new rays in his honour.[7]

Word about this new phenomenon spread like wildfire, inspiring admiration, fear and even poetry across the world. At the same time that Jules Verne's books about travelling to the centre of the Earth were capturing the public imagination, Röntgen had suddenly discovered the ability to see inside the human body. This led to some interesting misconceptions, like the idea that X-rays could see through a lady's clothing (the idea of seeing through men's clothing went unmentioned). The entrepreneurs of the time started selling X-ray-proof lead underwear, presumably only for women. 'X-ray glasses' were banned in a number of opera houses, despite not actually existing. Philosophers feared that X-rays could reveal a person's innermost self.

Hundreds of scientists around the world already had cathode ray tubes, a standard piece of equipment in physics labs. So, they first confirmed Röntgen's discovery, and then set about putting the tubes to work, all in a matter of months. Within a year of his discovery, in 1896, X-rays were being used to find bone fractures and shrapnel in soldiers' bodies on battlefields in the war between Italy and Abyssinia, and Glasgow Royal Infirmary had already set up the world's first hospital-based X-ray imaging unit.

In other areas of society, business people capitalised on the capabilities of X-rays for other services. Popular at the time was the 'pedoscope', which made X-ray images of clients' feet while they were trying on shoes, a practice later discontinued when evidence began to emerge that X-rays could sometimes cause damage to skin or tissue – an issue which we will return to later. Röntgen himself suggested another use by taking an image of metal weights inside an opaque box to show their potential use in industry. These early 'radiographs' paved the way for modern security scanners found in airports.

As he had decided not to patent his discovery and potentially hinder its medical application, Röntgen didn't see any income from all of this.

He wisely left the responsibility of developing these techniques to the medical profession, claiming to be too busy with his other research, but continuing to offer his assistance where it was needed.

Röntgen might seem a strange character: a 'lone genius' who made an 'accidental discovery' out of nowhere. After all, anyone lucky enough to have a phosphor screen nearby could have stumbled on the same discovery. But if we look a little closer, there were other factors at play. He had access to a large network of experts around the world, had many years of experimental training and had cultivated a practice of patience and humility even in the midst of his excitement. When he noticed the glowing screen, he had the knowledge to realise its significance and the curiosity to dig deeper.

Despite all the hype, no one really knew what X-rays *were*. Röntgen had shown that they didn't have quite the same reflection or refraction properties as visible light, or the ultraviolet or infrared light beyond the usual visible spectrum. There was no clear idea of how X-rays were created from cathode rays, or how they interacted with other matter, like the phosphor screen. His discovery had raised a whole swath of new questions about what matter and light were made of and how they interact. Answering these questions required further experiments with the cathode ray tube, which continued to play a central role in the discoveries that came next.

In early 1897 in Cambridge, England, Joseph John ('J. J.') Thomson, the founding director of the world's pre-eminent physical laboratory, aimed to settle a twenty-year-old controversy. Instead of focusing on the X-rays outside the tube, he wanted to determine the composition of the glowing cathode rays inside the tube.

Thomson had an unpopular hypothesis. He believed that the cathode rays were some kind of corpuscle, or particle. This put him at odds with Röntgen who, with his German peers, thought that cathode rays were immaterial, a form of light.[8] Thomson used the tubes available in his lab to study electricity in gases, but now he devised a new set of experiments designed to answer the question: what is the nature of cathode rays?

Thomson was the shy son of a Manchester bookseller, who announced at the age of eleven his intention to do original research. Where this precocious desire came from is unclear. His father passed away when Thomson was just sixteen, leaving no money for his education. Since no scholarships were available in physics, Thomson attended Trinity College, Cambridge, to study mathematics. There his quiet sense of humour – often expressed as a boyish grin – combined with his unshakeable intellectual self-confidence, frightened a number of his fellow students, who viewed him with almost a sense of awe.[9]

By the age of just twenty-seven, Thomson had been appointed as Professor and Director of the Cavendish Laboratory at Cambridge University. He was a rather short man with a straggling moustache, black hair which he parted down the middle and little care for his sense of style. An old friend later recalled how his bow tie would sometimes sit up by his ear while Thomson walked around, blissfully unaware. His home life was simple, but when it came to his speculations about the nature of matter and the Universe, he was quite the revolutionary.

Thomson started his investigations by carefully repeating the experiments of those who came before him. First, he wanted to establish that the cathode rays and the electrical charge they carry could not be separated. He used a magnet to bend the cathode rays, forcing them to hit an electroscope, a device for counting electric charge. A surprisingly large negative charge was recorded,[10] confirming his view that the rays really did carry electrical charge.

Next, he recreated an experiment attempting to bend the rays with an electric field, using a voltage held between two plates which his assistant had mounted inside a specially built vacuum tube. The rays, if they were the particles he thought they were, should be deflected by the voltage. On the other hand, if the rays were made of light, they ought to travel in a straight line and not be deflected at all, just as the light from a torch would pass by a voltage unhindered.

Thomson expected to see that the cathode rays would be deflected less by a smaller voltage than by a large one. Heinrich Hertz, a German physicist who had earlier discovered electromagnetic waves, had tried the same experiment before Thomson and found that while large voltages deflected the rays, smaller voltages appeared to have no effect. When Thomson first tried this in his lab, he was frustrated to get the same results as Hertz. It was as if the cathode rays were acting like particles for larger voltages, and like light for lower voltages. This was a big challenge to Thomson's particle hypothesis.

Thomson experimented with his apparatus, trying to understand what he was seeing. First he changed the type of gas in the tube, but there was no change in the result. Next he tried changing the amount of gas by pumping down the tube to a lower vacuum level, and found that the result changed: he saw small deflections with a small voltage, and large deflections with a large voltage, just as he expected. To be sure, he let some gas back in, and the small voltage deflections disappeared again. The small amount of gas left in the tube was becoming electrically charged and had the effect of cancelling out a small voltage, but not a larger one. As a result, the cathode rays simply didn't respond to a small voltage when gas was present. This turned out to be the cause of Hertz's results and Thomson's frustrations. As Thomson later wrote in his memoirs: 'The delicate instruments used in physical laboratories may, until their technique has been mastered, give one result one day and a contradictory one the next, and illustrate the truth of [the] saying that the law of the constancy of Nature was never learned in a physical laboratory.'[11]

All these results allowed Thomson to conclude that 'the path of the rays is independent of the nature of the gas'.[12] In other words, the effects he was demonstrating weren't due to the gas in the tube. Nor were they simply streams of charged gas molecules, as others claimed. They were much more fundamental than that. This led him to put forward his key argument: that all of these results were expected if the rays were indeed a type of negatively charged particle.

It remained only to show what *kind* of particle they were: atoms, molecules or something else. To determine this, Thomson used electric and magnetic fields to determine their charge and mass, in particular the ratio of the two, 'e/m'. It came out as a much larger number than he might have expected. It was a puzzling result and didn't fit with any known atom or molecule, which, as far anyone knew, were the smallest constituents of nature. Thomson had two possible explanations: either the particles were 'heavy', like atoms, with an extremely large negative electric charge, or they were very lightweight particles with a standard negative electric charge. Neither option was attractive. If the particles were atoms with a very large electric charge, he would have to rethink the idea of charge entirely. On the other hand, if the particles were light-weight, it would mean the atom was not an indivisible fundamental particle after all.

Thomson changed almost every variable he could think of, used different gases in the tube, tried different metals for the electrodes, and again varied the level of vacuum. Every version of the experiment produced the same type of new particle with the same large charge-to-mass ratio. He called on his knowledge of experiments from chemistry, observations of the light spectrum from stars, and even configurations of magnets in his speculations on the nature of the particles. Slowly but surely he leaned away from the idea of the particles being atoms with a very large charge. He was ready to announce his results.

On Friday 30 April 1897, barely a year after Röntgen had announced his discovery, Thomson emerged in evening dress in front of a packed crowd at the Royal Institution in London, and proceeded to re-create his series of experiments as a Friday Evening Discourse. These public lectures were held every Friday evening and drew enormous crowds of well-heeled Londoners[13] – the latest scientific discoveries were considered high culture in those days. At the climax of the lecture, he announced that the mysterious cathode rays were indeed a negatively charged type of particle, one which he determined was about 2,000 times lighter than hydrogen, the lightest atom. Thomson had discovered the electron, the first *sub*-atomic particle.[14]

It was an intellectual triumph. Thomson had dug into the mysterious glow of cathode rays and emerged with a new understanding of the nature of matter. By October of the same year, he was able to make a further leap: not only were cathode rays made of tiny particles, but those particles were a hitherto unknown component of matter that shattered the idea of atoms as the smallest indivisible entity. He was not yet sure where electrons came from, but believed that they were almost certainly held within atoms. Given the evidence, even Röntgen and his German peers had to concede that Thomson was correct. Between them, Röntgen and Thomson, using a single piece of equipment, had found two entirely new aspects of nature which had never been seen before.

We can now put their ideas together to explain what was happening inside the cathode ray tube. The high voltage across the cathode emits electrons at high velocity which are attracted to the positively charged anode. But some of the electrons, instead of hitting the anode, fly past it at high speed and crash into the gas and the glass wall, the energy transferred during this process creating light – the glow that had puzzled scientists for decades. This process is called 'bremsstrahlung', which translates to 'braking radiation', as the electrons brake inside the glass wall. If the electrons lose enough energy, they can create X-rays: a high-energy form of light – electromagnetic radiation – with the ability to penetrate hands (and other parts of the body).

Unlike X-rays, the utility of Thomson's discovery was not obvious at the time. Thomson openly wondered how such a small, inconsequential thing as an electron could ever be of interest outside of physics. In the early 1900s at the annual party of the Cavendish Laboratory, where he made the discovery, the proceedings included a tongue-in-cheek toast: *'to the electron, may it never be of use to anybody!'*[15] Yet twenty years after his electron discovery, Thomson presented another Friday lecture at the Royal Institution, this time on 'The Industrial Applications of Electrons', and with hindsight we can see that his discovery and our understanding of it would come to underpin the entire field of electronics.

How did this happen? It seems to make sense on the surface, of course, since electronics – as the name suggests – is dependent on the movement of electrons. But did Thomson's discovery have anything to do with it? Did we need his research, or would electronics have arisen anyway? To understand the relationship between Thomson's curiosity and the revolution of electronics, we need to be able to put his work in context.

At the Science Museum in London, there is a permanent gallery called 'Making the Modern World'. In a small, unassuming glass case in the middle of the walkway are a handful of glass objects with modest explanatory signs. One of these objects is an original cathode ray tube used by J. J. Thomson to discover the electron. In the same case is an early light bulb, and on the other side are two strange-looking objects called Fleming valves, which look like light bulbs with three pin-like legs on the bottom. This one display case is a history in miniature of the invention of early electronics.

Nearby sits a display highlighting another well-known inventor, Thomas Edison. In 1880, while scientists like Thomson were poring over cathode ray tubes in their labs, Edison and his assistants had stumbled on a similar technology in their attempts to make electric lights. At the time Edison was thirty-three years old, nine years older than Thomson, and he took a very different approach from the experimental scientists as he was driven by other motives – namely, making money from his inventions. Rather than exploring the detailed physics of the light bulbs, Edison's team simply tried as many materials and configurations as possible, a kind of 'cut-and-try' brute force tactic. Most types of filament would burn out almost immediately, but one member of his team, Lewis Latimer, an African-American inventor, had landed on a method of making light bulbs using a carbon filament which could last around fifteen hours.[16]

There was a problem, however: the glass surface of the bulb would blacken when it was operating, almost as if carbon particles were being 'carried' from the filament to the glass. Despite improving the vacuum as much as possible, the bulbs kept burning out. We now know this is because of the evaporation of the filament, but Edison didn't know that at the time. In one attempt to cure the problem, Edison tried to catch the carbon particles mid-flight by placing an extra electrode in the bulb and accidentally found that this caused an electrical current to flow, but only in one direction. It didn't solve the blackening problem, but the device seemed to be controlling the flow of electricity like a valve controls the flow of water. He called this the 'Edison effect'. He had no interest in *how* it controlled the flow of electrical current; he knew only that it did. Edison filed a patent for the 'Edison effect bulb' and then dropped the idea, as he couldn't see a use for it. He continued with his work on light bulbs, making small improvements to eventually extend the carbon filament lifetime to 600 hours, rendering the bulbs commercially viable. As for the 'Edison effect bulb', when someone later enquired as to how it worked, he told them he simply didn't have time to delve into the 'aesthetic' part of his work.[17]

Someone who did have time for the aesthetics – the principles underlying the work – was J. J. Thomson. In 1899, only two years after his original electron discovery, Thomson showed that just like the cathode ray tubes, the filaments in light bulbs were emitting electrons. Heating up the filament in the way Edison did was causing electrons to jump out in a process we now call 'thermionic emission'. This was quite different from the evaporating filament and turned out to be key to unlocking the Edison effect. Edison's seemingly useless invention sat unused for almost two decades until Thomson's work finally revealed how the extra electrode made current flow. When the electrode was positively charged it would attract the flow of electrons through the vacuum and complete the circuit, but when negatively charged it would repel the electrons and switch the current off. With this full understanding, Edison's 'valve' would find its use in a rapidly evolving world.

The next step in our story came in 1904, with the work of a consultant for Marconi's Wireless Telegraph Company, where radio and telecommunications were starting to emerge.[18] To make a telephone work, British physicist John Ambrose Fleming needed to convert a weak alternating current into a direct current. He had come across the Edison effect in 1889, when he was acting as a consultant to the Edison and Swan United Electric Light Company,[19] and realised that this was exactly what Edison's valve invention could do. The tiny signals emitted by radio transmissions would be enough to cause the valve current to switch on and off. The connection suddenly clicked in his mind, and he later wrote: 'To my delight I ... found that we had, in this peculiar kind of electric lamp, a solution ...'

The knowledge of the cathode ray tube was combined with the filamented light bulb, and the first 'thermionic diode', or 'Fleming valve', was invented: the first *electronic* device. Where *electrical* devices involved the flow of electrons through wires, *electronics* involved electrons moving through a vacuum, which could be controlled quickly and easily without the mechanical motion of earlier electrical devices. Fleming's invention caused a technological revolution. A few years later, an American inventor added a third electrode inside the thermionic diode, using Thomson's theories to guide him at every step.[20] By 1911 the 'triode' was being used as an amplifier and, soon after, streams of electrons in vacuum tubes were used as oscillators, as modulators of electrical signals and more. These purely electronic devices led to long-distance radio and telecommunications, radar and early computers. The electronics industry was born.

It's important to dwell a little on the two different approaches at play in this story. On the one hand, Thomson's curiosity-driven approach certainly seemed to be the key to understanding the workings of vacuum tubes, but he had no agenda to create anything other than knowledge. On the other hand, Edison's trial-and-error method led to entrepreneurial success, but he wasn't interested in developing a detailed understanding of how and why these technologies worked the way they did. Fleming was able to, in a sense, combine these two approaches and create a sophisticated technology. All the players were undoubtedly integral in the formation of the electronics industry, but none of it would have been possible without scientists constructing cathode ray tube experiments with no commercial purpose in mind.

The thing about seeking knowledge and understanding through the scientific process, rather than inventing a new product through trial and error, is that it tends to have a cumulative effect; it tends to develop more and more utility over time. This was true of the electron, and it was also true for X-rays because the two are linked. With the birth of the electronics industry, it was possible to make specific tubes for the production of X-rays, which created a flourishing market in X-ray tubes for medical and industrial uses. Examples of these tubes sit in the Science Museum gallery, just near J. J. Thomson's cathode ray tube and the early Fleming valves.

The rest of the story of X-rays plays out just a few steps away in the museum, in the form of a large medical machine made possible by the electronics industry and X-rays, a life-saving technology known as the Computed Tomography (CT) scanner.

Before the 1970s, if a patient needed to have a brain scan, doctors would perform something called 'pneumoencephalography'. A hole would be drilled into the base of the spine or directly into the skull and most of the cerebral-spinal fluid (CSF) would be drained from the patient. Air or helium would then be pumped in to create a bubble between the brain and the skull. The patient would be strapped into a seat that rotated in all directions, and they would be forced into different positions (i.e. upside down and sideways) to get the air bubble to move around in the brain and spine, while X-rays were taken in each position. The already sick patient would be forced to endure terrible pain, nausea and headaches, and they were often not anaesthetised during the procedure. All this was done just to get enough contrast in the

X-ray image to be able to tell the difference between the brain and the (now drained) cerebral fluid. After this torturous experience, doctors would examine the X-ray photographs and hope that they could tell if the shape of the brain was slightly distorted from lesions or growths within it. It was a brutal procedure. Yet it was the only option from 1919 until the 1970s.

At the time, X-rays only produced two-dimensional images. If you imagine the body as a box of liquid in which a series of objects (bones, organs and muscles) are located, it would be very difficult to see an object in the middle of the box with X-rays, as it would be obscured by everything in front of and behind it. Doctors find it difficult to understand 3D structures rendered in 2D. What was really needed was an innovation, one that could create a proper three-dimensional image.

In the 1960s Godfrey Hounsfield, an employee of EMI (Electric and Musical Industries), a large UK corporation that also dealt in electronics and other equipment, was looking for areas where computers could be useful and came up with a novel way of using them to make a better medical X-ray device. He had the idea of rotating the source and detector around the patient in order to take a series of X-ray images that could be digitally reconstructed using computers. This made possible the ability to create a full 3D picture of the inside of a body. It was called 'computed tomography', or CT.[21]

To make his idea a reality, he first built an experimental setup of a brain scanner. To test it, he went to local abattoirs and cut out brains from cows to take images of.[22] In an interview he described, with typical British understatement, that it was 'quite a job carrying [the brains] across London in a paper bag to put on the machine'.[23]

His early tests showed, in surprising clarity, a full 3D view inside organic tissue. The CT scanner even showed minute differences in tissue that Röntgen thought would never be possible: tissues in his early X-rays had been transparent, but combining many images meant that they now could be seen. It took computational power, a rotating setup and some clever mathematics, but the technique worked. The brain scanner was taken for trials at London's Atkinson Morley Hospital in 1971. It consisted of a specially designed movable bed on which the patient would lie, with their head positioned in a circular aperture which housed the scanning equipment – not too dissimilar, in fact, to how they look today.

The first patient to be scanned in 1971 was a woman with a suspected tumour in the left frontal lobe. The CT scan successfully identified the tumour, and as a result, it was operated on, restoring the patient to health. It was only then that Hounsfield and his team 'jumped up and down like footballers who had caught a winning goal'.[24] Finally, he realised the implication of his work: his invention had brought an end to the anguish of traditional cranial X-rays.

Hounsfield didn't stop with the brain scanner, which was announced to the world in 1972. He went on to build a machine that could reveal the inner workings of the rest of the human body, too. By 1973 the first CT scanners were being installed in hospitals in the United States and by 1980, and 3 million CT scans had been performed worldwide. Over time, CT scanners became so ubiquitous that by 2005, 68 million scans were being performed annually.

Since then, new ideas have led to real-time imaging, combinations with other imaging techniques (which we'll meet later) and the use of CT as a first-line technique in emergency departments. While it could take half an hour to collect an image in the 1970s, modern machines can do so in less than a second. Now, there are even CT techniques that help doctors navigate the heart in 3D during the insertion of stents, improving the success rate of the procedure. In addition, the internal structures imaged with CT can be printed using 3D printers to help doctors understand what is really happening inside a patient as they plan for surgery and implants, all without a single incision into the skin. The technology and capabilities continue to improve, focusing on increasing the scan speed, reducing radiation dose and taking ever more detailed 3D images.

The journey from X-ray discovery to modern CT scanners took more than seventy years to reach fruition. It required a series of inventions, breakthroughs in mathematical techniques and the dawn of computers all to coalesce. You can find some form of this technology in almost any hospital in the world. If you had asked doctors in Röntgen's day how to improve their knowledge of the inside of the human body, they might have just found a better scalpel. It was Röntgen and Thomson's quest to better understand a seemingly obscure area of physics that allowed their dis-coveries to provide an entirely *new* tool, which could then be perfected by Hounsfield and others to revolutionise medicine.

Of course, medicine is not the only area of society that has benefited from X-rays. Once you see them in use, you realise they are everywhere. Keep an eye out for the X-ray baggage-scanning machines next time you travel through an airport: they also had their origins in a lab in Würzburg. Beyond security applications, our material and physical world relies on our knowledge of X-rays. Companies who manufacture objects from oil pipes to aircraft, bridges to staircases, now use X-ray imaging to make sure that their products are up to standard. Where a crack develops or an air bubble exists, X-rays will reveal it, just like they would have in Röntgen's original experiments. This 'non-destructive testing' technique is a hidden part of the human-made world around us, but it's also the reason that our pipes rarely burst and aircraft rarely fall out of the sky. Non-destructive testing is a 13-billion-dollar industry which is constantly growing, and X-rays make up about 30 per cent of the market.

It took half a century for electronics and almost a full century for X-rays to realise their current potential, but even the story covered in this chapter is just a snapshot. The complete story stretches back through centuries of gradual accumulation of knowledge and technology, from the first laboratory vacuum created in 1643 by Evangelista Torricelli to the 1654 invention of the first vacuum pump by Otto von Geuricke. Expert glassblowers were needed to create the precise, yet delicate, apparatus with well-sealed joints to hold the vacuum. Equipment was needed that could provide voltages high enough to rip electrons out of the metal cathodes. The full process, then, spans many generations, even if it appears that the breakthrough happened in the blink of an eye.

It is simply astonishing how cathode ray tube experiments performed between 1895 and 1897 expanded our view of the electromagnetic spec-trum, blew apart the idea that atoms were the smallest particles in nature and led to the discovery of the first subatomic particle. If someone had been asked to predict the outcome of these experiments, they would have utterly failed to estimate their impact on our knowledge of physics.

But they would have been even *more* wrong if they'd been asked to predict the impact on society.

The other common thread in both Röntgen's and Thomson's discoveries is the fact that they were rapidly adopted into technology. Both ideas became integral to innovations in electronics and life-saving medical equipment that occurred over the following decades. Yet the fundamental concepts on which these technologies were based did not come from industry. They came from inquisitive minds performing experiments in an effort to increase our collective knowledge. Today many people associate the phrase 'cathode ray tube' with old-style televisions, but it is so much more than that. It represents the power that curiosity-driven research has to lead to groundbreaking innovation.

The experiments with cathode ray tubes overthrew the idea that physics was nearly finished. With the beginnings of *subatomic* physics, new vistas opened up for curious scientists. The next key experiments would come from one of Thomson's own students, when physicists started to ask: what else is inside the atom?

2

The Gold Foil Experiment: The Structure of the Atom

Ernest Rutherford had only been in Montreal a few months when he received an invitation to a debate from the local Physical Society. It was 1900 and the topic was 'the existence of bodies smaller than atoms'. Rutherford was eager to take part and wrote to his former advisor, J. J. Thomson, that he hoped to demolish his opponent, Frederick Soddy, an Oxford-trained chemist six years his junior. Soddy had always been intrigued by problems at the interface between physics and chemistry, but in Rutherford he would find a physicist who shook the very bedrock of chemistry.[1] The debate would kick-start one of the most astonishing series of discoveries in science and would lead not just scientists, but also artists, philosophers and historians, to completely rethink their assumptions about the world around them.

Soddy spoke first. He was a tall, serious-looking blond man with blue eyes. The youngest of seven brothers born in the south of England, as a schoolboy he overcame a speech impediment and turned his former nursery into a chemistry laboratory where he could carry out experiments, occasionally coming close to setting the house on fire. He had two firmly held values: truth and beauty.[2]

Soddy was there to defend the atom. His position was that the electron discovered by Thomson and others must be something different from the 'matter' that he and other chemists knew. 'Chemists will retain a belief and a reverence for atoms as yet concrete and permanent identities, if not immutable, certainly not yet transmuted,' he said. He threw down a challenge to Rutherford: 'possibly Professor Rutherford may be able to convince us that matter as known to him is really the same matter as known to us'.[3]

Rutherford stepped up to defend his position. Electrons, according to Rutherford, were part of ordinary matter. He described the work of Thomson and of those before him, people like Heinrich Hertz and Philipp Lenard in Germany, Jean Perrin in France and William Crookes in England. He reviewed Thomson's experiments to discover the electron and explained that, since electrons seem to come from matter, they must form part of the atom. Rutherford explained the new experimental results so well that he left the audience of McGill University students and staff convinced that they should change their long-held idea of atoms being the smallest immutable building blocks of matter. But though Rutherford may have won the debate, many questions remained about what was going on inside matter. The chemists and the physicists remained divided.

Rutherford – Ern to his friends – was a physicist, but he was about as far from the introverted physicist stereotype as you could imagine. He was tall, athletic and talked in such a loud voice that it would disrupt the sensitive scientific equipment in his lab. In frustration, his students would eventually build an elaborate light-up sign that hung over their experiments and said 'talk softly please'. According to science writer Richard P. Brennan, Rutherford had a 'deeply held belief that swearing at an experiment made it work better, and considering his results he might have been right'.[4]

Rutherford had arrived at McGill University looking slightly too young for his new role as its professor of physics, his career having been fast-tracked by the strong recommendation of his old advisor Thomson. Just a few years earlier Rutherford had moved from his native New Zealand to England and into a rising wave of new discovery in the field of radiation, diving in with the enthusiasm of a brilliant young mind out to prove himself. He quickly emerged as a star student at Cambridge, demonstrating independence in his research while his mentor was busy (although, to be fair, his mentor was discovering the electron at the time).

The discovery of radioactivity had happened somewhat accidentally in 1896, when French physicist Henri Becquerel was studying the light-emitting effects of uranium crystals. In 1898 Marie Curie discovered radiation being emitted from the element thorium, and with her husband Pierre joining her in her research, they announced the discovery of polonium[5] and radium and gave radioactivity its name all in one momentous year. During his graduate studies at Cambridge, Rutherford had joined the endeavour and demonstrated that there were at least two distinct types of radiation: alpha radiation, which could be stopped with a piece of paper, and beta radiation, which could be stopped by a block of wood.[6] Alpha, beta and, a few years later, gamma radiation were named using the first three letters in the Greek alphabet. At first their nature was unknown, although it wouldn't take long before Becquerel identified beta radiation as electrons in 1899 and Rutherford figured out that alpha radiation consisted of helium atoms which had lost two electrons – giving them a doubly positive electrical charge – in 1907. Though it was not known at the time, gamma radiation consisted of high-energy light, similar to X-rays. Rutherford's discoveries in radioactivity certainly caught Thomson's attention.

With his new professorship at McGill, his first research group and his own laboratory, Rutherford wanted to delve even further into the phenomenon of radioactivity. Canada provided a rather different atmosphere from Cambridge, but one that seemed to free him from the social constraints of an old English university so that he could do as he wished. He set his sights high: he wanted to understand the structure of the atom.

After their early debate in 1900, a genuine interest and collaboration emerged between Soddy and Rutherford, who each became ever more curious to understand the other's work. Soddy was so taken with learning more about radiation that he attended an advanced course Rutherford was teaching, learning about X-rays, radiations from uranium and thorium, and practical work on how to use an electrometer. As a chemist, he was most impressed by the electrometer, which could detect minuscule amounts of thorium based on the radiation it emitted. It was a much more sensitive method than simply weighing materials, as chemists were doing. In fact, the electrical method could detect a quantity of material 10^{12} (1,000,000,000,000) times smaller than their best analytical balance.

Meanwhile, Rutherford took on his first graduate student: a woman named Harriet Brooks. Women graduate students were exceedingly rare at the time, although perhaps Marie Curie's success had some influence. Brooks, the third of nine children in her family, came from a small town in western Ontario. Her father was a travelling salesman in the flour trade and there often wasn't much food to go around between the children. Disappointingly little is known of how she discovered her love for physics, nor of her personality or demeanour: these things were simply not written down.[7] What seems clear is that she sensed what higher education might bring her: the ability to escape her family home and become independent. After four years at McGill she emerged with first class honours and a number of scholarships won in mathematics and German, which relieved her family's need to support her. She was such a high-flying student that it was natural that Rutherford – who had no qualms about women doing research – had invited her to work with him.

Together, Brooks and Rutherford investigated the element thorium, finding that it gave off a mysterious 'emanation', a kind of gas that was like nothing they'd seen before. This was strange enough, but they also found that the emanation seemed to make nearby objects radioactive. That is, when the emanation was in contact with an object, it seemed to affect the object so that it spontaneously emitted alpha, beta or gamma radiation, in the same way that natural radioactive materials like radium and polonium did.

Brooks won a fellowship for her doctoral work with Rutherford and used it to travel from Canada to England in 1902 to work with J. J. Thomson, becoming the first woman to study at the Cavendish Laboratory. Based on her results, Rutherford started to think that someone handy with chemical techniques could help him understand what was happening, and invited Soddy to collaborate with him, who immediately abandoned his previous research work to accept.[8]

Soddy followed up on Brooks's work by using chemical methods, attempting to see if the thorium emanation would react with different chemical agents, but to no avail. He found that the temperature of the experiment made no difference, nor did putting the experiment in carbon dioxide instead of air. It seemed like the emanation was some kind of inert gas. He was sure that it was not thorium, but was created by the thorium somehow.

Finally, it clicked. The thorium was *turning into* the gas. Atoms of thorium were spontaneously changing form. It wasn't quite the alchemist's dream of turning lead into gold, but atoms were changing. Soddy was 'standing there transfixed as though stunned by the colossal import of the thing' and exclaimed 'Rutherford, this is transmutation!'[9]

We now know Rutherford and Soddy were observing the decay of radioactive elements, which turn into other elements by emitting alpha and beta particles, eventually forming stable substances. Nature had been doing alchemy for free all along. Soddy, who had just a few years earlier insisted that chemical atoms were immutable, had found evidence that completely overturned his worldview.

They went on to determine that radioactive decay follows an *exponential law*. There is a particular time taken, known as the 'half-life', for half of the atoms in a lump of radioactive material to change into another type of atom. If you start with a hundred atoms of oxygen-15 (a radioactive type of oxygen with an atomic mass fifteen times that of a hydrogen atom), then in two minutes only fifty atoms remain. The other fifty change to nitrogen-15. In a futher two minutes, there are just twenty-five atoms ($50 \div 2$). Another two minutes leaves 12.5 atoms, and so on. (You can't technically have half an atom, but the 'half-life' time of two minutes remains the same.) Matter was no longer the stable, unchanging substance it had always seemed.

Rutherford and Soddy's ideas were radical by the standards of the early twentieth century, so the scientific community had mixed reactions. In London the most senior figure in British physics, Lord Kelvin (William Thomson), simply refused to believe in atoms disintegrating. The chemists, who believed in the indestructibility of matter, were also up in arms about the implications of the work. At McGill, Rutherford's antics and theories of radioactivity also began to bother the other professors. The rest of the faculty thought his unorthodox ideas about matter might bring the university into disrepute: members of the Physical Society where he and Soddy had debated were extremely critical and advised him to delay publication and be more cautious.[10] At one point, his fellow professors pulled him into a meeting and told him in no uncertain terms to tone things down. Rutherford stormed out of the room, barely able to disguise his temper.

He wouldn't toe the line for long. In 1904, while walking around campus, he came across the geology professor Frank Dawson Adams. Without any preamble, he asked Adams how old the Earth was supposed to be. Adams hazarded a number of 100 million years, based on various methods of estimation at the time. Rutherford thrust his hand in his pocket, pulled out a black rock and said, 'Adams, I know beyond any doubt that this piece of pitchblende in my hand is seven hundred million years old', and then walked off.

The constantly decaying radioactive matter in nature, Rutherford had realised, could be used to estimate the age of the Earth. Rocks contained a small amount of the radioactive atoms he and Soddy had been studying. If he knew the rate of decay from one atom to another, he could count the number of un-decayed atoms compared to the number of 'daughter' particles and calculate how long the object had existed. Rutherford had come up with the idea of 'radiometric dating'. His first estimates were based on uranium-238, where '238' refers to the atomic mass number. Elements with different mass number are called *isotopes* and can have different radioactive properties, despite being the same chemical element (Soddy discovered isotopes and invented the term in 1913). Uranium-238 has a half-life of 4.51 billion years and slowly decays though a series of intermediate steps to form lead-206, which is stable. With rough estimates of the half-lives from his lab, Rutherford had compared the quantity of uranium and lead in a sample of pitchblende and discovered that it was much older than the supposed age of the Earth.

It's one thing to show off to geology professors, but he still had to convince the physicists and chemists that he was right about the transmutation of atoms. Rutherford travelled to England, where on 20 May 1904 he gave a speech at the Royal Institution and presented his discoveries about radioactivity. In the audience, he spotted Lord Kelvin. Kelvin was already struggling with the idea of atoms disintegrating, and Rutherford knew the last part of his speech, when he planned to talk about the age of the Earth, would be especially tough for him. Kelvin was considered the authority on the age of our planet, based on a calculation he had performed of the Earth's rate of cooling.[11] Rutherford recalled, 'To my relief, Kelvin fell fast asleep, but as I came to the important point, I saw the old bird sit up, open an eye and cock a baleful glance at me! Then a sudden inspiration came, and I said Lord Kelvin had limited the age of the Earth, provided no new source [of energy] was discovered. That prophetic utterance refers to what we are now considering tonight, radium! Behold! The old boy beamed upon me.'[12]

As evidence arrived from other laboratories, it confirmed the idea that many elements were unstable and had half-lives. Lord Kelvin publicly abandoned his previous position against radioactivity at a meeting of the British Association for the Advancement of Science, and was forced to pay out a bet to another physicist, Lord Rayleigh, as a result. The rest of the community slowly accepted that radioactivity really did happen the way Rutherford and Soddy had surmised.

When Rutherford was awarded the 1908 Nobel Prize in chemistry, he remarked that he'd witnessed many transformations in his laboratory, but none as quick as his sudden transformation from physicist to chemist. Soddy also went on to win the Nobel in 1921 for his contributions to radiochemistry, nominated by Rutherford. As for Harriet Brooks, she was in Cambridge when Soddy and Rutherford made their transmutation discovery in 1902, but J. J. Thomson was too preoccupied to notice her work. She later returned to Canada in 1903 and continued producing excellent research in radioactivity until she got engaged in 1905, at which point the college she taught at told her she'd have to leave her job if she got married.[13] She broke off her engagement and continued to work. In 1907, after meeting and then working with Madame Curie in Paris, Brooks faced a tough choice. Another Canadian professor, her former lab demonstrator, had started making romantic approaches to her through a series of letters. She was thirty-one years old and the societal pressure to marry and have children was intense. Rutherford – by then in Manchester – tried his best to ensure she would be financially independent by attempting to hire her. In his reference letter he attested that she was the most prominent woman physicist in radioactivity next to Curie. In the end, Brooks chose to accept the offer of marriage and moved back to Canada, where she had three children. Her career in physics ended. It was only recognised in the 1980s that her work was integral to Rutherford and Soddy's discovery that elements disintegrate and transform into other elements.[14]

For most people the Nobel would be the peak of their career, but for Rutherford it would turn out to be just the first step. He still hadn't answered his original question: what is the structure of the atom? His ability to make imaginative leaps and follow them up with simple but powerful experiments had made his name. In 1907 he returned to the UK to lead the physics department at Manchester University. What he discovered next would require physicists and chemists to make an even bigger leap in thinking, based on one of the simplest, but most famous, investigations in physics: the gold foil experiment.

Despite the many advances that Rutherford had made, the experiments he was building by 1908 were still very primitive. He described his approach best himself: 'we have no money, so we shall have to think'. The students and staff in Rutherford's research group were famous for using things like tin cans, tobacco boxes and sealing wax, combined with a lot of hard work. There was joy to be found in figuring out how to test nature using unsophisticated but smart methods. One of his students, Australian physicist Mark Oliphant, would later write, 'He was full of ideas but they were always simple ideas. He liked to use words to describe what was going on.'[15] It was the same with his view of the atom.

Rutherford described his idea of the atom at the turn of the twentieth century as 'a nice hard fellow, red or grey in colour, according to taste'. It's easy to imagine that the tiny atoms that make up our food, our bodies and our planet are like little billiard balls, a picture we're often taught in school.[16] In 1908, even though it had been ten years since Thomson discovered the electron, physicists still had no real idea what the internal structure of the atom was. But Rutherford was beginning to get an inkling that the composition of the atom and radioactivity were intimately linked.

Thomson's view, and that of many others, was that the atom was a sphere of positive charge, with negatively charged electrons embedded within it – dubbed the 'plum pudding model'. There were a few other ideas around, such as the 'Saturnian' model put forward by Japanese physicist Hantaro Nagaoka, of 'a central attracting mass surrounded by rings of rotating electrons', but there was no evidence to suggest that this model was in any way accurate.[17] Rutherford held Thomson in very high regard, but he was beginning to doubt his old advisor.

Rutherford's domain was growing, as were his responsibilities. He was now overseeing a whole department at Manchester, housed in an impressive red-brick building with purpose-built laboratories and offices. One of these Rutherford set aside for himself as his personal laboratory. Like many of the other labs it had heavy wooden floorboards and the walls were covered in tiles: mustard yellow tiles near the floor, a deep red stripe at desk height, then cream tiles extending all the way to the ceiling. It may have felt austere, but it was an eminently practical kind of place. Here, Rutherford could set to work exploring the question of what the atom really looked like inside. Or rather, his staff and students could.

As the lab director, Rutherford would have been too busy to conduct most of the experiments with his own hands, even if he had wanted to. Instead his job was to gather a team of people who would all work together towards the laboratory's goals, while he would drop by on his rounds to see results, give suggestions and keep up motivation. In one of these lab rounds, Rutherford met Ernest 'Ernie' Marsden, a twenty-year-old undergraduate student from Lancashire who was a bundle of energy and enthusiasm. Rutherford towered over the slightly short Marsden, as he did over everyone. The son of a cotton weaver, Marsden grew up adoring music and literature as well as science and chose to study physics after the influence of teachers in high school. He was prone to infectious laughter and was, according to his colleagues, always a joy to be around.[18] Marsden was in need of a research project for his undergraduate thesis. Rutherford threw out an idea.

Back in Canada, Rutherford had observed that when he passed alpha particles through a thin piece of metal they formed a fuzzy image on a photographic plate. If the metal sheet was taken away, the image in the photograph was clear. The alpha particles seemed to be scattered, perhaps deflected by the atoms in the metal, but he didn't know why. It was a tiny effect, one that most people would probably pass over. Rutherford encouraged Marsden to do experiments to check the effect in more detail.

To advise Marsden, Rutherford set him to work under the supervision of Hans Geiger, a German-born physicist six years his senior. Geiger was born in Neustadt, in the Rheinland Palatinate, a beautiful wine district. He was fascinated by the natural world, and took pleasure and pride in building experiments. He had recently completed his doctorate and moved to Manchester when Rutherford arrived. Later, he would become well known for inventing the eponymous Geiger counter. Rutherford offered up his own personal laboratory to the two younger men for their experiments.

Members of Rutherford's group had already studied how electrons scatter when they pass through metals. They found that the electrons would undergo a series of collisions with the metal atoms, and a few electrons would be deflected back in the direction they arrived. Now the question was how alpha particles would behave in a similar experiment. Alpha particles (or helium nuclei, as we now know them) are around 7,000 times heavier than electrons and this heavyweight status means a large force is needed to make them change course as they travel along. Intuitively, they should pass straight through a thin piece of metal. Yet the fact that Rutherford had observed alpha particles forming a fuzzy image when he sent them through the metal sheet was quite interesting. The questions were clear: if they fired alpha particles one by one at metal, how would the thickness of the metal affect the way they scattered and deflected?

It was good training for Marsden to help set up an experiment and this one was fairly typical for Rutherford's lab. These kinds of experiments involved hour upon hour of staring through a microscope at a screen and counting little flashes of light from alpha particles. It would take time and stamina, so Geiger and Marsden got to work.

The experiment relied on a variation of a vacuum tube. Rather than construct a cathode ray tube, which produced electrons, they instead wanted to use alpha particles. In one end of the tube they placed a strong radioactive alpha source made of radium and then sealed the other end with a window of mica, a thin material through which alpha particles could travel. They angled the tube at 45 degrees onto a thick piece of metal and on the reflected 45-degree angle placed a zinc-sulphide detector screen which would give off a flash of light if an alpha particle hit it. They made sure to place lead between the alpha emitting tube and the detector, to prevent stray alpha particles from travelling directly to the detector and upsetting the results. The setup was designed so that only alpha particles which were truly reflected off the metal would be picked up. Geiger and Marsden then got in position to watch for the flashes on the screen.

First, they observed what happened when alpha particles hit the surface of a thick piece of metal. Just as with the electrons, a few of the alpha particles were reflected. For thick sheets of metal, the alphas were so far behaving roughly like electrons. Inside the metal, the deflection of the alpha particles from each individual atom was expected to be small. A thick sheet of metal contains many layers of atoms, and although the alpha particles are 7,000 times heavier than electrons, the result confirmed their prediction that even these heavy projectiles could sometimes turn back on themselves after enough collisions. Did the type of metal make a difference? It appeared it did. Metals made of heavier elements like gold reflected more alpha particles than light ones like aluminium.

Next, Geiger and Marsden checked whether the thickness of the metal made a difference. They reasoned that if they made the metal foil thin enough, the alpha particles would all travel straight through, but might scatter a little, as Rutherford had observed. They chose gold for this part of the experiment because it could easily be made into thin foils. They gradually changed the thickness of the gold foil and checked how many 'scintillations' they observed on the screen. As the foil thickness was reduced, the alpha particles appeared to start travelling straight through, as expected. But then they noticed something odd: no matter how thin they made the gold foil, the zinc-sulfide screen still occasionally lit up. Around one in every 8,000 alpha particles appeared to bounce off the foil and hit the screen. This wasn't just a nudge that altered the alpha particle direction slightly; this was a huge effect that completely deflected the alpha particles and sent them towards the screen as if they were reflected from the foil. But how could that be? There was nothing they knew of inside the atoms of gold which could cause this effect. It seemed to defy all known laws of physics. How could a heavy alpha particle be deflected by tiny electrons, or the diffuse positive charge of the atom?

Geiger and Marsden brought the news to Rutherford. He later described it as 'quite the most incredible event that has ever happened to me in my life. It was almost as incredible as if you fired a 15-inch shell at a piece of tissue paper and it came back and hit you.' After hearing the results, Rutherford had to think of every plausible explanation that could fit the data and rule them out, one by one. If the plum pudding model was correct, the deflection of the alpha particles would only be small, but that wasn't what Geiger and Marsden had observed. They had to figure out how the alpha particles had been turned back on themselves, which would have required an enormous force to be present in the gold atoms. There were a few options to consider: the experiment might have been wrong; the alpha particles might somehow be absorbed and re-emitted by the atoms; or possibly, all the positive charge of the atom might be condensed in a central area inside the atom.

The experiment was conducted in 1907–1908 and published in 1909, but Rutherford's theory on what it meant for the atom took until 1911 to come to fruition. During this time, Rutherford went away and did the calculations, taking the time to enrol on a course in mathematics to be sure he got it right. He continually found that there was only one explanation that fitted the data: atoms *must* be composed mostly of empty space with a tiny, dense nucleus.

If Rutherford wanted to overturn the accepted model of the atom, he would have to demonstrate beyond doubt that the new model was correct. Over the next few years, aided by Geiger's invention of a counter for measuring, Marsden and Geiger carried out another series of experiments in which all the pieces fell into place. Only after this did Rutherford put his new theory out into the world. The atom wasn't a plum pudding dotted with negative electrons: it had at its core a tiny nucleus of positive charge, dense enough that it could deflect alpha particles when they came close. The electrons were part of atoms too, but they were orbiting the nucleus at an enormous distance. If the atom was the size of a cathedral with the electrons at its walls, the nucleus was the size of a fly. In between was nothing at all.

Geiger and Marsden's experiments completely changed the view of atoms and in turn, the Universe. Far from being the solid entities they had been viewed as for millennia, atoms were mostly made of empty space. It's hard to over-estimate how much of a surprise this result was. As Arthur Eddington wrote in 1928:

When we compare the universe as it is now supposed to be with the universe as we had ordinarily preconceived it, the most arresting change is not the rearrangement of space and time by Einstein but the dissolution of all that we regard as most solid into tiny specks floating in void. That gives an abrupt jar to those who think that things are more or less what they seem. The revelation by modern physics of the void within the atom is more disturbing than the revelation by astronomy of the immense void of interstellar space.[19]

It might just seem an interesting detail to understand what the inside of an atom looks like. Yet this discovery and understanding the mechanisms for radioactive decay and transmutation came to dominate science, technology and even politics for decades. The fact that atoms are made of a tiny, dense, positively charged nucleus, surrounded by the negatively charged electrons, gave birth to the whole field of 'nuclear physics'.

From such simple experiments came the possibility of an immense amount of knowledge. It made Rutherford so excited that on one occasion, C. P. Snow, a chemist and one of Rutherford's collaborators in Cambridge who later became a well-known writer, recalls Rutherford bubbling over and shouting out at a meeting of the British Association 'We are living in the great age of science!' while the others in the room sat in stunned silence.

His enthusiasm was warranted: he saw the potential that might come from understanding the atomic nucleus and how radioactivity works. Today many people associate the word 'nuclear' and 'radioactivity' with the technologies of nuclear power and nuclear weapons which emerged decades after these discoveries. The power unleashed by our exploration of the nucleus and by the invisible nature of radioactivity can sometimes incite fear. Yet if radioactivity did not exist, if all the elements were stable, if the nucleus were not wonderfully complex, we along with our planet and everything on it simply wouldn't be here. Radioactivity happens because the atom has a structure, and the discovery of that structure led us to a deeper, more fundamental understanding of the nature of matter, sought after for millenia.

Radioactivity is a natural process. It embodies the idea that everything in our lives, even matter itself, is in an ever-evolving state of change.

This change is excruciatingly slow in some cases, so we call some atoms 'stable', by which we mean we haven't yet seen them decay, since their half-lives are much longer than the age of the Universe. But other atoms are decidedly unstable. They have half-lives from billions of years down to days or minutes and for that reason they are much more interesting – and often useful – to us.

These radioactive elements are found naturally in rocks, in the air, almost everywhere. The granite of your kitchen worktop can contain uranium, thorium and their radioactive decay products. Some elements like potassium (chemical symbol K) have both stable and unstable isotopes, and differ in atomic mass because their nuclei have varying numbers of neutrons, which can be more or less than the number of protons. Isotopes of the same element can have different radioactive properties. For instance, most potassium is the stable isotope K-39, but 0.0012 per cent is the isotope K-40, which has one extra neutron, which emits mostly beta radiation (electrons) with a half-life of 1.3 billion years. This means that even bananas are, technically, radioactive. The radiation dose is, however, minuscule, and you'd have to eat 5 million bananas in one sitting to feel its ill effects. Our own bodies contain these isotopes too, inescapably. We are all radioactive.

Today we rely on naturally occurring radioactive elements for many technologies, from smoke detectors (where an americium source of alpha particles causes a small current, which is tripped if smoke scatters the alpha particles away) to radioactive sources lowered down deep holes and used to map out the composition of the surrounding rock. This technique, known as 'borehole logging', stimulates gamma ray emission from the elements in the rock and lets the user assess with minimal digging whether there are precious minerals, oil, gas or other valuable commodities buried deep down under the ground. Other radioactive sources have been used for many years in cancer treatment, and to sterilise mail, particularly after anthrax poisoning attempts were made in 2001 by transmission through the post, after which US government mail has been sterilised using radiation.[20]

The use of natural radioactivity in other areas of society is so much a part of our world that it's easy to forget it didn't exist before the discoveries of Rutherford, Soddy, Brooks, Geiger and Marsden. One need look no further than the museum located just a few paces away from Rutherford's old lab, the Manchester Museum, for proof of this. It doesn't contain any old physics equipment, but does have plenty of fossils (including a huge T. rex skeleton called Stan). There's a reproduction of an enormous tree root system from the upper Carboniferous period, with a sign indicating it is 290–323 million years old. There's a plesiosaur that was found in North Yorkshire by a group of university students, its 180-million-year-old fossilised bones laid out on the floor in a huge glass case. It's easy to assume we've always had techniques to determine the absolute age of fossils, rocks and ancient artefacts, but as Rutherford's interaction with the geologist Professor Adams reminds us, we haven't. The main reason we know objectively the age of pretty much any unrecorded historical object is because of our knowledge of radioactivity.

After Rutherford's discovery of the nucleus, it took time for physicists to understand nuclear physics well enough to know why the half-lives of different atoms vary. In the meantime, the discovery of many unstable atoms with different half-lives in nature gave us a large range of tools and techniques to put dates not just to fossils, but to almost anything. It's impossible to list all the things we know about because of radiometric dating techniques, but consider these few.

We know the Shroud of Turin is a medieval forgery[21] and we can put a date to the Dead Sea Scrolls. We know Homo sapiens migrated out of Africa not once, but over multiple periods,[22] and we know how they spread across the globe because we can date human remains, like the 14,300-year-old ones found in a cave in Oregon.[23] In archaeology, we can put a timescale to objects not just locally, but compare them over different countries and even continents, letting us build a story of the prehistory of the world. We can date ice as far back as 1.5 million years old[24] to understand the ancient climate from ice cores. It's also thanks to radiometric dating that we know when dinosaurs roamed the Earth *and* the date of the asteroid which appears to have wiped them all out 65 million years ago.[25] Going further back, we can identify the first evidence of fossils that might be animals, a kind of early sea sponge found in 665-million-year-old rocks in the Trezona Formation of South Australia.[26]

This knowledge forms such a rich part of the cultural and historical context of our lives and our species. We can put all those stories together accurately not just because we can compare rock strata and skeletons to one another, but because atoms spontaneously decay into other atoms: because Rutherford, his team and other scientists after him developed and refined these methods. The quest to understand the smallest objects in nature might have seemed like an obscure bit of physics at the time, but it has come to underpin much of our understanding of culture, art, geology and our place in world history.

Once again it was simple experiments, conducted by a few individuals, which led to groundbreaking new knowledge: that there was a tiny nucleus at the heart of matter itself. This discovery also presented a plethora of questions that were important to follow up on. How was the nucleus held together? How did the electrons stay in the atom? The first answers to these questions would come from the early days of quantum mechanics, born of experiments designed to study the nature of light and its interactions with matter. In time, physics would grow into a field of increasing complexity and the simple experiments that Rutherford was so fond of would cease to reveal the secrets of the atom. Even the radioactive substances found in nature would not prove potent or flexible enough and would eventually become the limitation, rather than the tool, of discovery. Technological and theoretical progress would start to move hand in hand with experiment. Physicists would start to make surprising connections between seemingly disparate aspects of nature. Now, our story takes us to the first of those surprises, where the interactions between light and matter led physicists to adopt a startling new view of our world at its most fundamental level.

3

The Photoelectric Effect: The Light Quantum

What is light? Debate about the nature of light has raged since the seventeenth century. At first, light was supposed to be like a particle,[1] an object travelling at speed through the hypothetical aether in straight lines – an idea championed by Isaac Newton. On the opposing side was Dutch physicist Christiaan Huygens, a major figure in the scientific revolution who discovered Saturn's moon Titan and then established the mathematics of the wave theory of light in his 1690 'Treatise on Light'. Huygens argued that light was a wave vibrating its way through the aether (which would later turn out not to exist[2]). Although Newton's great prestige forced the particle theory to dominate for a long time, experiments as always trumped scientific celebrity and one model came out on top: the wave theory.

The main experiment that settled the debate for the wave theory was first carried out by Thomas Young in England in 1801. The modern version is easy enough to recreate, and most physics students have a go at it. It begins with a laser pointer, directed onto a black metal plate with two tiny slits cut into it. This gives the experiment its name: the 'double slit experiment'. Behind the double slit is a viewing screen. The question: what will we see on the screen? Our intuition jumps to an analogous experience. Imagine a fence in the sunshine with two slats missing: the fence blocks the sunlight and casts a shadow on the pavement, but two bright patches appear in the gaps where the slats are missing. Most people think the laser light will produce two bright red lines of light on the screen, the double slit being the equivalent of the two missing slats, with the rest of the screen in shadow. That is what we'd expect – but that's not what happens. Instead, a set of *interference fringes* emerge: bands of light and dark patches spreading out across the width of the screen.[3]

This *interference* is a unique property of waves. For instance, we can make a similar pattern using water waves. If you head out to a still pond carrying two inflatable balls, hold one ball in each hand about a metre apart and bob the balls up and down in sync to create two sets of waves, you can see that a pattern[4] emerges. Where peaks of the two waves meet each other, they cause 'constructive' interference; the opposite case, where peaks and troughs overlap, leads to 'destructive' interference and the waves cancel each other out. The result is a beautiful fan-shaped pattern alternating between areas of waves and stillness, which spreads out away from you across the pond.

The interference effects of light also show up in our everyday lives, but in ways subtler than shadows. They produce the spectacular colours on a soap bubble, the iridescent colour of butterfly wings or the rainbow colours you see when looking at the back of a CD or DVD. Interference looks a little more complex in these situations because they use white light (composed of many colours as opposed to a single colour laser pointer), and interference patterns depend on colour, so these scenarios produce colourful patterns instead of bright and dark patches.

Young's double slit experiment shows this interference in action: at some positions on the screen, light added to light gives even brighter light and at others, light added to light gives darkness. By measuring the distance between the bright spots on the screen and knowing the wavelength of the light from the laser pointer, we can use the wave theory of light to predict what will be observed. Once nineteenth-century scientists added this knowledge to the evidence that light can *diffract* and *refract* as well as *interfere*, all of which are properties of waves rather than particles, the debate was settled: light is a wave.

Throughout the nineteenth century the classical wave theory of light went from strength to strength, predicting all the known behaviour of light observed in the lab. Based on it, we could build and understand microscopes and telescopes, mirrors and lenses. We could explain how rainbows worked, why the sky is blue, and many other phenomena. The classical theory held up well even after Scottish physicist James Clerk Maxwell connected it with his theories of electromagnetism, which gave us a superior definition of the nature of light waves. To be more precise, we can say that light is an *electromagnetic* wave travelling at a speed of almost 300 million metres per second, which we call c. The wave has an oscillating *electric* component and a *magnetic* component, which swap back and forth. By the year 1900, the nature of light was no longer in question.

Then a series of experiments began to cast serious doubt on the wave theory. They showed that light wasn't always acting like a wave – because sometimes, light seemed to act like a *particle*. Problems began to crop up for the classical theory when scientists started to ask how the wave theory interacted with other parts of physics. Questions which had previously been swept under the carpet came to the forefront. Why should *light* and *matter* be treated as different from each other? What makes light act one way and matter act another? As physicists pondered these questions, the radical idea that neither light nor matter are quite what we thought they were began to emerge. This marked the beginning of a complete revolution in physics and the start of the peculiar yet wonderful theories of *quantum mechanics*.

Let's take stock of where our journey has taken us since finding X-rays in Röntgen's lab in 1896. Electrons and the gold foil experiment showed physicists that atoms aren't the smallest things in nature, since the tiny electrons which carry negative electric charge are inside atoms. Atoms turned out not to be the stable, everlasting entities that the chemists wanted them to be: instead, physics revealed that atoms can change, transmuting into different elements by emitting radiation, altering form repeatedly until they reach a point of stability. Atoms were no longer solid spheres of matter: instead, they consisted mostly of empty space. All these revelations foreshadowed the next major discoveries which changed physics almost beyond recognition. We even have a different name when we teach the physics that emerged after the turn of the twentieth century. We call it *modern physics* as opposed to *classical physics*, as if everything that came before the theories of this era was just a little ordinary.

The foundation of the problem was laid in 1887, when Heinrich Hertz surpassed his earlier discovery of electromagnetic waves by accidentally discovering that light can make sparks. Put more precisely, he found that if you shine ultraviolet light onto a metal surface, electrons are ejected. This connection between light and electricity is called the *photoelectric effect* and became a popular research topic,[5] with many physicists including Wilhelm Hallwachs and Philipp Lenard in Germany, Augusto Righi in Italy, J. J. Thomson in England and Aleksandr Stoletov in Russia all trying to understand how it worked.

Light, according to the wave theory, carries a certain amount of energy proportional to its *amplitude squared* (the size of the wave, or brightness of the light). The physicists studying the photoelectric effect suspected the electrons in the metal were bound up in atoms, so an electron would need to gain a little bit of energy to get bumped out of the atom. Having overcome this initial energy barrier, applying more and more light should transfer more and more energy to the electron, until it darts out with an energy corresponding to the absorbed energy from the light (minus the energy required to escape the metal). From this, three predictions could be made. One, brighter light should lead to electrons that travel faster. They reasoned that the stronger the light that shone on the metal, the *more* energy the electron would have and thus the faster it would travel away from the metal. It seemed reasonable. Two, if the light is dim enough, there will be a little bit of a delay as the energy needed to escape slowly accumulates, after which the electron would leave the metal with a low speed. Three, because electrons needed to jiggle around and absorb energy to escape, the temperature of the metal should make a difference to the result.

In 1902 Philipp Lenard, a Hungarian-born physicist working in Germany,[6] had found that there was a problem with the very first prediction: he saw no correlation between the speed of the electrons being ejected and the light intensity. Lenard even came up with a hypothesis that the whole idea was wrongheaded – that light energy is not transformed into electron energy in the photoelectric effect at all and that instead, the light was just a trigger telling the atoms to release electrons.[7] This 'triggering' hypothesis seemed unlikely, but no other compelling explanation existed.

*

On the other side of the world, another experimental physicist was playing catch-up. Robert Millikan, an associate professor at the University of Chicago, was determined to make his mark in the field of physics but was plagued by a lack of equipment and the fact that no one else in his laboratory was interested in what he was doing.

Millikan first discovered his love for physics after his Greek teacher at Oberlin College in Ohio asked him to teach a course in the subject. Despite having no prior knowledge of it, he taught himself physics one summer by doing every problem he could get his hands on in the textbooks. He went on to do a PhD at Columbia University and then spent a year studying in Germany before taking up a position at the University of Chicago. Millikan was well known for his incredibly rigorous schedule: he worked twelve hours a day, six hours teaching and six hours of research.

It was fortuitous that the year he spent in Germany was 1895–6, when the great discoveries of X-rays and radioactivity were made, as this helped him form new ideas for his research. But in Chicago, despite his intense schedule and his unfading optimism, he felt a keen lack of progress in his research due to his isolation. Millikan knew that Lenard would be producing results in Germany surrounded by other experts, while he was working almostly entirely independently.

Like all labs at the time, his looked very different from modern ones. This was the early 1900s after all: electric lighting was new and not very efficient, so the lab was more like a dull factory than the bright white spaces of today. Most houses in the Chicago area would still have been lit using gas lamps or candles, as they would have no electricity for another twenty years. There were no computers, of course. All calculations were done using slide rules, pencil and paper, and there was no company Millikan could call upon to construct his scientific equipment, meaning everything had to be built in-house. For these reasons, it took a lot of dedication to decide to embark on an experimental research problem at the time, but that was what Millikan wanted to do.

All he needed was a good problem to get to work on. For that, he would have to read all the latest research papers, which was useful because he was also responsible for organising the weekly seminar series for his department. In a bid to liven up discussions one week, he brought along and presented a research paper that impressed him greatly, one that we've already encountered: J. J. Thomson's 1897 paper on the discovery of the electron. Millikan was so inspired by Thomson's work that he decided this was the new and thrilling topic he would work on. He wanted to study electrical discharge in high vacuum, but there were no vacuum pumps in the laboratory that would do the job.

Back then, vacuum pumps were mostly mercury pumps: elaborate yet delicate concoctions of interconnected glass tubes and bulbs all hand-made by glassblowers. Liquid mercury would be pushed through and as it went it would remove a few molecules of air. If you repeated that enough times, eventually you'd remove enough air to achieve a good vacuum. But Millikan had to start from scratch, and over the course of three painstaking years he tried and failed repeatedly until he ultimately devised an even better apparatus. To the standard mercury pump he added a tube that contained charcoal immersed in liquid air. By 1903 he managed to evacuate enough air that his experiment had a billion times lower pressure than the atmosphere[8] – a very respectable vacuum level even today. He was ready to make some measurements.

While Millikan had been wrangling with vacuum pumps, a new book came out from J. J. Thomson[9] that laid out the prediction that photoelectric emissions were supposed to depend largely upon temperature, as all experimentalists had found up to that date.[10] According to the classical view, at a higher temperature the electrons in the metal should have *more energy*, so they would be released from the metal much more easily and at a higher speed than from a metal at a cold temperature.

Armed with the high-vacuum setup, replicating these results sounded like a good starting point to Millikan. He shone light onto a metal electrode, which was temperature-controlled within the glass apparatus. To measure the speed of the electrons, like other experimenters before him, he would apply a voltage which the liberated electrons had to fight against; higher-speed electrons would need a higher voltage to be stopped. But when Millikan tried the experiment with his vacuum system, he found the results to be *completely independent* of temperature. What had he done wrong?

Millikan divided up the problem and set some of his graduate students on it. They worked together in a small room, stepping over trays of sulphuric acid and calcium chloride that they laid out to dehumidify the air to prevent water building up on the electrodes in their experiments. It took three or four days of continuously blowing clean air over the system before they could take a reliable measurement, and weeks would go by when they would be plagued by problems of air leaking into the vacuum system and would have to start again from scratch. Despite the challenges, Millikan eventually managed to take an aluminium electrode all the way from 15 to 300 degrees Celsius and measure the electric current emitted. Again, no temperature dependence was found. Their detailed work continued for years, with the team building a complex in-vacuum setup with a movable wheel to which they attached eleven different metallic discs: copper, nickel, iron, zinc, silver, magnesium, lead, antimony, gold, aluminium and brass. The wheel sat on agate bearings inside an 8cm-diameter glass cylinder, with a narrow light source – smaller than each disc – shining down the tube. They placed a strip of iron on the edge of the wheel so that carefully waving a large magnet near the tube would rotate each metal sample into the path of the light source without having to open the system to air.[11] What they found was that *all the results* were independent of temperature, at least up to 100 degrees Celsius, which was the highest they dared go with the eleven-disc version of the equipment. Millikan later wrote that so far he 'seemed to be having very little success as an experimental physicist!'[12]

But Millikan's results were actually a success. Because they were different from those who came before him, he had developed that most elusive and precious of scientific states: a gap in knowledge. He must have had a hunch that his lack of positive results might be indicative of something greater than a simple experimental mistake. After all, he'd spent *years* making sure the experiments worked reliably. So what was the alternative explanation? If his results were correct and the photoelectric effect really wasn't dependent on temperature, then classical physics would be simply unable to explain the photoelectric effect at all.

In Bern, in 1905, Albert Einstein would come across the photoelectric effect and make a theoretical leap that would help guide Millikan's experiments. Einstein had studied physics in Zurich, where in the evenings he would continue working with his fiancée, Mileva Marić, a

Serbian-born physicist and the only woman on his course.[13] After his final exam Einstein had been unable to find an assistant job in physics, so he temporarily left to take up a poorly paid teaching position in Winterthur, 20km to the north. One day in 1901 he wrote to Mileva saying that he was 'filled with such happiness and joy'.[14] She might have expected him to be happy, as she'd just written telling him he was about to become a father. But the reason he was so excited was that he'd just come across Lenard's experimental results on the photoelectric effect, showing that electrons could be produced by ultraviolet light.

Einstein considered it strange that most areas of physics were particle-like: atoms, electrons and the vibrations of individual molecules that caused heat all relied on the motion of individual, *discrete* objects. Even water waves consisted of small objects – molecules of water – in collective motion, while sound waves were pressure waves in molecules of gas. Yet light waves were considered a *continuous* phenomenon. Why was that?

Einstein was aware of recent work by his older colleague, German physicist Max Planck, a fan of deep, fundamental theoretical physics. When he was young Planck had chosen physics over music despite his physics professor telling him that 'almost everything is already discovered, and all that remains is to fill a few holes'. Planck had recently come up with a fascinating new idea, an attempt to bring together different areas of physics, combining mechanical vibration (heat) and electromagnetism (light). Planck had started out recognising that there was definitely some relationship between heat and light: objects emit different colours of light at different temperatures, so hot coals glow red, while the light from the Sun is more yellow or white.

When I say light, I don't just mean the visible spectrum. Light, or more properly *electromagnetic radiation*, stretches in frequency from X-rays and gamma rays all the way to infrared and radio waves. But for our purposes, I'll just refer to this as *light*. So why do objects glow a particular colour? What stops hot coals from glowing purple or the planet Jupiter from glowing in X-rays?[15] Once again, classical physics came up short.

Previous physicists had tried to determine the light which would be emitted by a sort of simplified hot object called a *blackbody*, an imaginary entity introduced in 1859 to help in understanding how heat radiates. A blackbody is what would form if you took a box – or cavity – and held it at a constant temperature. Over time it would produce a unique kind of light called *blackbody radiation*.[16] The key thing about blackbody radiation is that it doesn't matter if the blackbody is as small as a pea, or as big as a planet – as long as it both absorbs and emits radiation perfectly, the *spectrum* of light it produces, that is the amount of light of each colour emitted from a blackbody, is always the same. This is what makes it unique. Experiments trying to approximate blackbody radiators showed that the amount of light emitted always increased with frequency at first, peaked at some colour and then decreased again at high frequency. The peak depended only on the temperature of the object. This can be seen at a blacksmith's forge where metals first glow red, then orange, then eventually white as they get hotter, the peak of the spectrum shifting from red towards blue.

Using classical physics to calculate the light emitted from a blackbody resulted in an equation that didn't match the experiments at all. These earlier calculations, by British physicist Lord Rayleigh, predicted that the amount of light emitted at the lower (redder) end of the spectrum would be small, but then moving through yellow, green, into blue, purple and ultraviolet the amount would grow and grow, eventually peaking with high-energy X-rays and even higher-frequency gamma rays. For every doubling in frequency, the amount of light emitted should increase by a factor of four. But this clearly wasn't right: when we look at the world it's not entirely blue and purple,[17] or cooking us with high-energy X-rays. The prediction was also impossible because if you add up the total amount of light power radiated across all frequencies the total would be *infinite*. If this were true, then all of matter, even the coldest matter, would radiate so intensely that all the energy would disappear in a puff of high-frequency light. This was such a problematic state of affairs for theoretical physics that it became known as the 'ultraviolet catastrophe'.

Planck could not accept this situation. On taking up the problem in around 1900,[18] he realised that in these earlier calculations of the spectrum of emitted radiation, some assumptions were made about how energy was behaving inside the blackbody. It had been assumed that the energy could be shared between atoms (or 'resonators') in the box in any way at all, so that there were infinitely many ways in which the energy could be distributed.[19] But this meant that when adding up the total emitted power, *all* these possible states contributed and had to be added together, which is why the power came out as infinite. Planck realised he could use a mathematical trick to escape the problem, but he didn't like it.

If energy could only be absorbed or emitted in chunks, that is, if energy had some smallest size, then the number of ways it could be shared would no longer be infinite.[20] This is just like the limited number of ways in which you can divide a group of ten people – I can have five people in my group and you can also have five, or I might have ten and you zero, or perhaps four and six, but it would make no sense to talk about having 2.32 and 7.68 people in the groups. That's because people are discrete rather than continuous objects.

Planck treated the problem as if energy came in discrete packets, and mathematically speaking this helped him avoid the problem. To put in such a trick, Planck introduced the smallest bundle of energy that could be transported, which he called a *quantum* of energy. What's more, for his maths to work, he had to define that energy could only come in integer multiples of this basic amount. The size of this energy amount was tiny and was related to the frequency of the light through a new physical constant that Planck invented, h, which he said had a value of around 6×10^{-34} J.s.[21] He saw no other option that would produce the right result, but since 'a theoretical explanation had to be supplied at all costs, whatever the price'[22] he made what he called an 'act of desperation' to resolve the issue.

Planck didn't really think energy came in little packets, but he found that his mathematical trick worked. This method produced an equation where the amount of light emitted by a blackbody first increased, then peaked at some colour and then decreased again at higher frequencies. Most importantly, his equation matched the experimental data. But while his method worked, his results didn't cause a revolution among physicists. The new radiation law was accepted quickly, though the fact that he'd had to invoke the very odd idea of energy quantisation to achieve it was ignored for the most part.[23]

Einstein took Planck's idea seriously, however. He decided to believe that energy really did come in little packets and took it one step further. He proposed that light itself consisted not of waves but of these very same little bundles of energy: *quanta*. By pushing the idea well beyond what Planck had intended, Einstein said that light itself is discrete; light is made of what we would now call *photons*. He then came up with a theory which could explain the mysterious photoelectric effect.

His theory said that a photon would give up all its energy to a single electron in the metal. The energy of the photon was just the frequency (colour) multiplied by the constant that Planck had earlier come up with, *h*. He predicted that if someone did an experiment where they changed the frequency of the light and measured the energy of the photo-electrons, the results would lie along a straight line, the slope of which would give *h*. Brighter light should give *more* electrons, but the energy of them would only depend on the *frequency* of the light. The theory also made a second prediction that below a certain cut-off frequency, it wouldn't matter how bright the light was: no electrons would be released at all, as the energy arriving from the light wouldn't be high enough for the electrons to escape the metal. Forget the temperature, he was saying: look at the frequency.

When he published his paper in 1905, no one had yet done a detailed investigation of the relationship between energy and frequency that would demonstrate whether Einstein's theory was right. But over in Chicago, there was a frustrated experimentalist who had the experience, the ambition – and now the equipment – to try it out.

Robert Millikan didn't believe Einstein's theory, which isn't really surprising considering the theory wasn't very well received in general. Even Max Planck didn't take it very seriously, despite being the originator of the idea of the quantum and the editor of the journal that accepted Einstein's paper for publication. Planck clearly thought Einstein's idea was a bit far-fetched and would later comment in a recommendation letter: 'sometimes, as for instance in his hypothesis of light quanta, he may have gone overboard in his speculation [but it] should not be held against him too much.'[24] For Millikan, though, it wasn't just an amusing idea. He really thought that Einstein's theory had to be wrong, because light was so obviously a *wave* rather than a *particle*. He believed that the hypothesis that light was made of quanta was a 'bold, not to say ... reckless hypothesis'. It went against the clear evidence, like the double slit experiment we discussed earlier, of light being a wave-like phenomenon. How could it possibly be made of particles?

Now, Einstein had made a clear prediction against which to compare experimental results, and Millikan saw the chance to make his name as a physicist. Back in the lab and with a renewed vigour, in 1907 Millikan set out to disprove Einstein.

By now he and his team had become meticulous, eliminating any source of possible error in the equipment. They still used the same basic setup: a source of light, a metal surface and something to count electrons, but it became ever more elaborate. He'd changed from using spark-gap light sources – where high-voltage electrodes cause sparks through a gas to produce light including in the ultraviolet – to more stable ones, because sparks were creating electrical oscillations and introducing errors. He also found that to get reliable results the surface of the metal had to be very clean, otherwise they might be measuring the photoelectric effect of some surface build-up of metal oxide instead of the pure metal. Eventually, by 1909,[25] Millikan's team came up with a system that included a sharp knife which spun inside the vacuum system and scraped off the surface of the metal before it was exposed to light. Each time they shone the light on the metal surface, they would measure the energy of electrons that were emitted by applying an electric field sufficient to stop them.

From the start of this enterprise to the point where Millikan published his final results took twelve years. In that time there would have been a string of research students arriving, working and graduating from his lab. He made two major experimental campaigns in 1909 and 1912, and it took him until 1916 to publish the results. Millikan's earliest experiments in 1903 had already confirmed that the photoelectric effect didn't rely on temperature at all. After Einstein had made his predictions, Millikan came back to the problem, figuring he'd be able to show that such preposterous ideas as quanta of light weren't needed and that a simple tweak to the classical wave theory would be sufficient to explain the experimental data. The persistence he showed in his attempt to prove Einstein wrong probably seems as if it borders on obsession, and we might ask what took him so long to get there. The reason is an endearingly human one: Millikan's results frustrated and plagued him because he was trying to *disprove* Einstein, yet his experiments showed he was doing nothing of the sort.

In fact, Millikan found that almost every prediction of Einstein's turned out to be supported. The energy of the emitted electrons was directly proportional to the frequency of the incoming light, just as Einstein said. He showed that below a certain cut-off frequency, no electrons were measured at all, as would happen if light were made of quanta. He even measured Planck's constant, h, to within 0.5 per cent, by far the best measurement to that date. Millikan had found the best evidence yet that the theory he'd set out to disprove was, in fact, correct.

At the end of his 1916 paper, Millikan made it clear that while he accepted the experimental results, he still just couldn't believe the implications of what he'd found. It would be easy to think that even if Millikan were grappling with this new theory, his results would lead all the other physicists to suddenly accept Einstein's view that light was made of quanta, but they didn't. Millikan confirmed that the predictions made by Einstein's theory were correct, but no one had really seen a particle of light, so most scientists were perfectly happy to just ignore the concept of light being a particle and treat the photoelectric effect as unsolved.

They were avoiding a seemingly ugly and counter-intuitive concept: if you take into account both Millikan's results showing that light acts like a stream of particles and the centuries of evidence showing that light acts like a wave, the conclusion has to be that light has both a particle nature *and* a wave nature.

As William Henry Bragg, a British-Australian physicist, joked at the time, quantum theorists 'describe light as a wave on Mondays, Tuesdays and Wednesdays and as a particle on Thursdays, Fridays and Saturdays'. But however we describe it, we just have to accept reality as we find it. Sometimes it's possible to have such a strong intuitive mental image of nature that it leads us to get stuck in a mindset that something must be either A (a wave), or B (a particle). But in this case, in some situations we can use A, the wave theory, and in other situations we can use B, the particle theory. Neither is wrong, and which one is applicable depends on how we choose to do our experiments.

One point to clarify is the question of how Young's double slit experiment works, given the particle nature of light. If we do Young's double slit experiment with just one photon at a time, what happens? Even in that situation, each individual photon will act like a wave, and if you wait until enough single photons build up a pattern on the screen, you'll see the same interference pattern as with the more powerful laser pointer. Each individual photon seems to somehow go through *both* slits. This is fine if you think about light as a wave, but confusing if you think about it as a particle.

The ins and outs of the philosophy of quantum mechanics would be a whole different book, but what matters is how nature really behaves, which is what experimentalists aim to figure out. This is why science is ultimately an experimental subject, because no matter how good a theoretical model is, no matter what 'facts' we think we know, at the end of the day we're constantly trying to describe something that happens in nature, which can only be probed using experiments. While Einstein's description of light as a particle was a fascinating theory, it was Robert Millikan who painstakingly gathered the evidence that nature really behaves this way, yet few have ever heard of him.

Explaining the photoelectric effect was so important that Einstein was awarded the Nobel Prize in 1921 for his work on it – not for his better-known theories of relativity. Two years later, in 1923, Robert Millikan[26] was also awarded the Nobel Prize. By the time of his acceptance speech, he had changed the backstory a little bit, claiming that he'd set out to verify Einstein's theory and to measure Planck's constant all along. Actually, it had taken both him and the rest of the physics community that long just to accept what his results really showed.

Today, quantum mechanics is the best description we have of reality on the smallest scales and is far from just obscure philosophy. The eventual theory that emerged and which fully describes the particle-like and wave-like nature of light is now called Quantum Electrodynamics (QED) and took another forty years after Millikan's experiments to really come to fruition. QED incorporates both quantum mechanics and Einstein's special relativity, and we'll come back to it in more detail later. For now, the important thing about QED is that we can use it to calculate quantities in nature to a precision of better than one part in a *billion*. Scientists in many fields and high-tech industries now use quantum mechanics in one form or another on a daily basis, and we all use the results of it in our daily lives, without even realising it. Not having an answer to why nature acts this way (and we really can't answer why) doesn't mean we can't learn it and use it.

The idea that Millikan was studying – which we now understand in terms of light imparting energy to electrons both in a vacuum and within materials – is not just something that happened once in a laboratory experiment and faded into obscurity once it was understood. Quite the opposite.

Millikan would have worked at his desk with paper and pencil, and on hot days he would have opened the window to get some air flowing. Today we work with laptops and can simply use a remote control to switch the air conditioner on. Inside that remote control there is an LED (light-emitting diode) which sends a binary signal in an invisible (infrared) light. When we press the button, the photons from the remote travel out and hit a detecting photodiode placed in the air conditioner and – just like in Millikan's experiments – these photons free up electrons, giving them kinetic energy. The photodiode is made of a material called a *semiconductor*, which can be used in two layers. This forms a junction that lets electricity flow more easily in one direction than the other, so photodiodes let electricity flow when light is shining on them.[27] The air conditioner responds to the electrical signal it receives, interpreting the binary pattern and following our command. That binary pattern differs between your TV and the air conditioner, which is how they manage not to confuse each other. To someone in Millikan's day, all of this would have seemed like pure magic.

The properties of semiconductor materials combined with the physics of the photoelectric effect enabled the development of a vast array of electrical components starting in around the 1940s, which are now produced in enormous quantities worldwide. Solar (or *photovoltaic*) cells are a type of photodiode that converts photons from the Sun into electrical current efficiently enough to power homes and businesses. They have enabled some phenomenal human endeavours, like satellite communications and space exploration, but they aren't the only application. A lot of the seemingly effortless adaptation of technology around us is the work of these little photodiodes.

All those sensors that turn on the lights when you enter a room, that dispense soap and hand sanitiser, that open doors for you: they all use proximity sensors, which reflect infrared light off an object (you) and back to the photodiode. The closer something is, the more light will be reflected, which creates an electrical current. It's the same technology that's used in most security systems.

The reason photoelectric-based devices are so useful is because they output a current that's proportional to the amount of light falling on them – as long as the frequency is sufficient to produce electrons, more light produces more electrons and thus more current. This means the output is linear and works well with all our other electrical and electronic components. For instance, GPS sports watches now use photodiodes in light-based heart-rate monitors to continuously read the wearer's pulse through the wrist. A green light shines on the skin and with each cardiac cycle the amount of light reflected from blood flowing in subcutaneous tissue changes, the photodiode picks up the changes in this light and an algorithm calculates and displays the heart rate.[28] Your smartphone senses if it's bright or dark outside and auto-adjusts the screen brightness depending on the amount of light falling on it. The same tech is used to automatically change car dashboards from day to night mode and controls the aperture and shutter speed on a modern digital camera.

The indirect applications of photodiodes are probably even more numerous. All laser-based measurements use them, which means that almost every road and building around you would have needed them in the surveying and alignment process. Photodiodes are also used to pick up light-based signals in communications networks using optical fibres; if you've got fibre broadband, that network uses photodiodes to convert signals from light back into electrical pulses to transmit information to you from around the world. Our speedometers and odometers use them, and the feedback systems which keep motors in electric vehicles running smoothly use them too. Photodiodes are used to control the positions, speeds and operations of many automated processes in factories, so unless it's completely hand-crafted, photodiodes were used in the manufacturing of most of the things you own.

All of these things are evidence of our understanding of the photoelectric effect, and would not be possible without our underlying knowledge of fundamental physics that grew from these first foundational experiments. Millikan's research – together with the double slit experiment and the data on blackbody radiation – gave physicists a firm foothold on which to build their new quantum mechanical view of reality. When quantum mechanics took hold, its application quickly extended beyond explaining light. Quantum mechanics is the theory that describes all of matter too.

After Einstein and Planck's contributions, many other physicists joined in the development of quantum mechanics. With every new problem that cropped up in physics, quantum mechanics would evolve and figure out how to solve it. When it came to the nature of matter, this was essential. Rutherford's model of the atom – the tiny nucleus and orbiting electron model from Chapter 2 – seemed untenable when physicists realised it should be unstable: the electrons should emit radiation, collapsing in on the nucleus in a light-emitting death spiral. But Niels Bohr, a young Danish theoretical physicist, solved that by using the idea of *quantisation* to explain how electrons are arranged around the nucleus. Electrons can only possess certain values of energy – their energy is also quantised – which means that they orbit at distances from the nucleus depending on their energy values.[29] Electrons can move up or down between energy levels by either absorbing or emitting radiation in the form of light (a photon), but they can't sit between these levels. There also exists a minimum value of energy an electron can possess, which is the closest an electron can get to the nucleus.

It wasn't until 1923 that a French aristocrat, the younger son of the duc de Broglie, picked up where Einstein left off in questioning why physics treated *light* and *matter* differently. In his PhD thesis Louis de Broglie noted that quantum physics seemed to agree that light could act like particles, but in that case, was the opposite true? Could material particles act like waves? The answer, it turns out, is yes. *Any* particle or piece of matter, whether massive like a proton or massless like a photon, also has a wave nature, and the relationship between the energy and the frequency of the wave is $E=hf$, where h is (once again) Planck's constant. The theory that emerged, wave quantum mechanics, could describe all kinds of new behaviour of atoms and particles. It even told us that the subatomic particles aren't solid entities, but merely have a certain probability of being found in a particular state or location at any given time.

This idea of matter being made of waves is hard to believe. When you lie down, you don't fall through the floor; if somebody hits you, it hurts; if you accidentally try to walk through a clear glass door you discover, embarrassingly, that you can't. All of this would lead you to believe that your body is a solid object and that the matter it is made from is a continuous, unbroken surface. Yet you are made almost entirely of nothing. Even with the earlier view that matter is made of solid particles – where the nucleus and the electrons have some definite size – the volume of actual matter in each atom is so small that if you took all the matter from every human on Earth and lumped it together, you could fit it into a space no larger than a sugar cube. Now we can see it's not even that simple, because 'matter' is not perfectly solid. With the advent of quantum mechanics, everything changed.

These new ideas caused a stir not just in physics but throughout society. Their effects on the public imagination were felt keenly by the artist Wassily Kandinsky, who wrote:

> The crumbling of the atom was to my soul like the crumbling of the whole world. Suddenly the heaviest walls toppled. Everything became uncertain, tottering and weak. I would not have been surprised if a stone had dissolved in the air in front of me and became invisible. Science seemed to be destroyed.[30]

Matter is not certain or deterministic, but is associated with probabilities and waves. The solidity of matter turns out to be just a consequence of the interactions between wave-like entities. Electron waves pushing back against other electron waves are keeping you hovering ever so slightly above the surface you're sitting or standing on right now. As far as we know, everything that happens in the world, and in your body and mind, emerges from these small-scale interactions. It brings a whole new perspective towards your fellow humans.

If this muddles your sense of reality, you are not alone. You are experiencing what Millikan, Kandinsky, Planck, Rutherford, Bohr and even Einstein struggled to accept. We're not conscious of the wave particle nature of matter because we're not able to interact with matter in a way that would let us notice it in an everyday sense. We're human-scaled, not quantum-scaled. We don't see wave-like qualities of everyday objects, because the wavelengths are so small that we can't measure them. The *de Broglie wavelength* is inversely related to an object's momentum – its mass multiplied by its velocity – so once something has the mass and energy of a cricket ball thrown at 160km/hr, its wavelength shrinks to just a billionth of a billionth of a billionth of a micrometre (which we write as a decimal place followed by thirty-three zeroes then a one, or in scientific notation 1×10^{-34} m). By the time we get to the scale of people, the wavelengths are even smaller: a Usain Bolt-like object sprinting the 100 metres has a wavelength 200 times smaller than the cricket ball, around 5×10^{-37} m.[31] These wavelengths are far too small for us to notice any funny wave-like behaviour, so we can just use classical physics to approximate their movement and get away with it. But we can't do that when we get down to very small objects like atoms and particles, and on this scale *all* of the experiments conducted since its discovery tell us that quantum mechanics is right.

But have we ever seen the wave-like nature of particles? Absolutely, yes. In 1925, shortly after de Broglie's work, American physicists Clinton Davisson and Lester Germer of Western Electric (later Bell Labs) performed the first experiment bouncing electrons (which had a wavelength of roughly a nanometre) off a nickel crystalline structure in a metal and demonstrated that electrons form interference patterns just like light waves. A molecule just nanometres across has a de Broglie wavelength under 1 picometre (a thousandth of a nanometre), and those have been made to interfere as well. There's something of a competition among physicists for the largest object to show interference in a double slit experiment. The current record holder is Sandra Eibenberger, who conducted a masterful experiment during her PhD in 2013 in Vienna and observed interference in giant molecules containing 800 atoms, which contain more than 10,000 individual subatomic particles.[32] On this scale, the molecule wavelength is about 500 femtometres, around 10,000 times smaller than the molecule itself. Researchers are now asking if they could even create interference patterns with living biological objects such as viruses or bacteria, which would allow them to start asking fascinating questions about whether consciousness will ruin the wave-nature of the experiment, or if living organisms can also be in two places at the same time as they travel through the double slit apparatus. They think it will take them about ten years to get the experiment to work.

One key point of wave particle duality, and one which even physicists sometimes get confused over, is whether a *single* electron displays interference with itself, just like single photons in the double slit experiment. The answer, of course, is yes. By the time those experiments were carried out in the 1970s everyone assumed it had been done before. An Italian team led by Giulio Pozzi in Bologna and a Japanese team led by Akira Tonomura[33] at Hitachi (the two experiments were independent) didn't even publish their findings in a physics journal, choosing an education journal instead.[34] As they had already accepted that particles acted like waves they didn't think they were showing anything new at all. It was just that, by the 1970s, the teams had the equipment that enabled them to do the experiment, a device which relies on wave particle duality to function, and one which turns out to be more common than most of us realise: an electron microscope.

Electron microscopes were first invented in the 1930s, but nowadays you can buy one of these devices for around \$2–3 million from a high-tech supplier. How common are they? It's very hard to say, but my estimate is that there are tens of thousands in labs, companies and research institutes all over the world. Across campus from my lab in physics at the University of Melbourne, the biology institute called Bio21 is home to a number of these devices.

The building contains clean, bright workspaces populated by scientists in white coats, but the electron microscope lab presents quite a contrast to the racks of conical flasks and test tubes, sinks and fume hoods. A cylindrical metal device a few metres high accompanied by electronic racks, the electron microscope sits in a dedicated room. A green beam flits around on a fluorescent screen through a viewing window, which allows users to look in as the microscope operates. There is a single computer which controls the device and allows users to see images just as a regular light-based microscope would.

The many different researchers who use these microscopes are united by their need to see tiny objects and how they interact, down to the atomic scale. Unfortunately, this is beyond the capability of normal light-based microscopes, which can only measure things down to a limit – known as the *resolution* – of about 200 nanometres, or a magnification of about 2,000x. For biological molecules and even some electronic components which are smaller than this, a regular microscope produces a fuzzy image, because it's only possible to see things that are the same size or larger than the wavelength of the light being used.

Using an electron microscope, researchers can capitalise on the fact that particles also have a wavelength – the de Broglie wavelength – and the higher the electron energy is, the smaller the wavelength. This allows electron microscopes to work at wavelengths down to picometres, able to resolve objects down to a nanometre – a *billionth* of a metre – or even less. The ability to see at this scale has enabled an explosion in applications of 'nanotechnology' since the late 1980s, allowing scientists and engineers to study and build atom-by-atom structures and compounds used for everything from textiles to food manufacturing through to drug design.

Quantum mechanics and wave particle duality aren't just important for microscopes or for physicists studying atoms; they have powerful direct implications on chemistry and biology too. Quantum mechanics has a direct influence on how molecules form, interact and bond: this is the motivation for research in quantum chemistry. In biology, many fundamental processes for life are quantum-mechanical in nature. The new field of quantum biology is only just catching up with the many ways in which classical physics fails to explain these processes, and the breadth of processes which require quantum mechanics to explain them is remarkable: from photosynthesis, to how birds navigate during migration.

I mentioned earlier that some of the electronic components based on semiconductors use the photoelectric effect directly, but actually that downplays the importance of quantum mechanics in electronics. *All* modern electronic devices use our understanding of quantum mechanics. The evolution of computers from the early days of vacuum tubes, which we encountered at the start of the book, to the transistors and chips that are in every modern phone, computer, car and appliance, have their basis in quantum effects. Specifically, they rely on the way that wave-like electrons in silicon can only take on certain values of energy, so create 'energy levels' – in a similar way to the electrons around the atom – but when you put lots of atoms together in a crystal-like formation, this can

THE MATTER OF EVERYTHING

change the allowable levels.[35] Because we now understand the physics of this, we can manipulate the properties of silicon very precisely using techniques we'll meet later in this book. The quantum mechanical nature of light and matter has also enabled us to create lasers, atomic clocks (which are crucial to our GPS navigation systems), and many other technologies that we rely on every day. Our world would be unrecognisable without the applications of this theory.

Our future technologies are likely to be almost entirely based in quantum mechanics. Quantum computing is rapidly growing in utility, which is why the physics department at Melbourne University also has a large electron microscope in the basement. It's used to take images of thin layers of diamond on silicon, into which physicists carefully implant helium ions in a process called 'doping'. Physicists are using these techniques to make quantum devices that can be used as the basis for quantum computers. The electron microscope, a technology that came out of early understanding of quantum mechanics, is being used to make the next generation of quantum-based technology, continuing the research and technology feedback loop.

In this chapter, we saw how the problems that cropped up for classical physics ended up leading to an entirely new description of nature on small scales: quantum mechanics. In the midst of it all, Robert Millikan and his team spent twelve frustrating years in the laboratory perfecting their craft to gather that first crucial information about the details of the photo-electric effect, eventually showing that Einstein's reckless hypothesis was correct. Millikan didn't invent quantum mechanics, but his experiments were important in establishing that quantum mechanical theories genuinely reflected the reality of nature. This is how knowledge progresses. There is no sudden moment of inspiration, but we inch our way along in the dark anyway, often spending long stretches feeling out details, trying to establish a hold on some corner of this Universe we live in. Eventually, it clicks, and a new image of nature starts to take shape in our minds.

Today we celebrate quantum mechanics as a theoretical and conceptual triumph, and this is undoubtedly true, but without experiments we would never have known that quantum mechanics actually describes the behaviour of the world around us. We would never have been able to use it – in a practical sense – as we now do. From these detailed and obscure experiments grew our understanding of the subatomic world. That knowledge has played a large role in creating electronic devices, computers, solar panels, and instruments with the ability to image objects at scales that light microscopes can't reach – all of which rely on the strange consequences of the subatomic world not behaving according to classical physics.

So far we have seen how a handful of experiments broke down classical physics, unseated the idea of the atom as the smallest piece of matter and led to a totally new view of physics in which atoms – consisting of mostly empty space – can change over time, light can act like a particle and particles can act like waves. X-rays and the electron, radioactivity and the atomic nucleus, and now quantum mechanics changed our world for ever. But there were more surprises in store. Having spent the last few chapters exploring deep inside the heart of matter, it is time to lift our eyes upwards. Our attention turns to surprises about nature that quite literally rained down on scientists from above.

Part 2

Matter Beyond Atoms

While our thirst for knowledge may be unquenchable because of the immensity of the unknown, the activity itself leaves behind a growing treasure of knowledge that is retained and kept in store by every civilisation as part and parcel of its world.

–Hannah Arendt, *The Life of the Mind*, 1973

4

Cloud Chambers: Cosmic Rays and a Shower of New Particles

On the hill above the Hollywood sign sits a majestic white stone building that looks out over Los Angeles. It is not a mansion, but a public museum: the Griffith Observatory. There visitors watch planetarium shows and view the night sky through telescopes, exploring their place in the cosmos. Inside among the cool, dark marble is a series of exhibits, one of which – in a metre-squared Perspex box – holds the key to the next step in our journey. It is inconspicuous, somewhat overshadowed by chunks of meteorite, moon rocks and an enormous image of the night sky. But curious visitors are rewarded with a mesmerising experience: tiny condensation trails form sporadically against the black background, around twenty of them every second. They emerge suddenly, descend gracefully for half a second, then disappear.

The device is a *cloud chamber*, an early type of particle detector that lets humans see particles as they travel past in a 100-millionth of a second. Inside are short tracks as thick as a pencil formed by alpha particles (helium nuclei) and thin, light, spider-web-like tracks which are mostly electrons (beta radiation) or gamma rays. These entities are smaller than atoms, objects in nature that we have no way of seeing, touching or otherwise detecting with our own senses. Yet here is a device that lets us see them. Though we cannot perceive these particles directly – they are far too small for that – cloud chambers instead let us see the effect they leave behind.

This particular version in the Griffith Observatory, called a diffusion cloud chamber, is a 1936 design by American physicist Alexander Langsdorf, perfected from the original invention of the early 1900s. It is a simple idea, but one that transformed our understanding of the fundamental constituents of nature. Inside the sealed chamber a vapour of alcohol begins at the top and then falls down to a cold metal plate at the bottom. As it falls and cools it enters a state called supersaturation, in which any tiny disturbance will cause the vapour to form droplets. If any charged particles zip through they ionise the vapour, leaving just enough energy in their wake to form a tiny streak of cloud, like the white contrail we see behind a jet aeroplane.

In this chapter we follow the cloud chamber from humble beginnings to its heyday in the early 1930s, when it facilitated a series of remarkable discoveries – including completely unexpected new particles that changed our view of matter. These were particles that were not, in fact, parts of atoms at all. We'll see how this new particle detector led experimentalists beyond their basements and up into the mountains, opening up new vistas and leaving theorists racing to catch up. We'll also see how these new discoveries about matter led to entirely new ways of seeing inside pyramids and volcanoes, and revealed new insights about our planet.

This new era of discovery started with a seemingly simple question, the same one that visitors to the Griffith Observatory often ask if they take the time to observe the unceasing shower of particle tracks travelling through the cloud chamber. That is: where are all the particles coming from?

In the early 1900s, it was almost exactly this question that scientists were asking. In their words, they wanted to find out where the extra radiation they saw in their instruments was coming from. Studies into radiation were performed at laboratories in Berlin, Vienna and Cambridge using a simple and rather crude device called an electroscope. One property that was easy to predict was the so-called inverse square law, whereby if the experimenter stood twice as far away from a radiation source, the level detected would drop by a factor of four. At least, it was supposed to behave like that, but some astute scientists had noticed that their instruments appeared to pick up some extra radiation. Why was there more radiation than they expected? Without the answer to that, they could barely hope to understand what was happening in their experiments in the lab.

The answer seemed simple: it was coming from minerals in the Earth. In her work to discover radium and polonium – both used as laboratory sources – Marie Curie famously spent years working in a shack milling and refining tonnes of a mineral called pitchblende. These two new elements were a precious commodity for scientists studying the properties of radiation, and they originated from the Earth itself. So, logically, it must have been these same minerals creating the troubling background radiation. The answer seemed clear and so did the way to test it. If the radiation came from Earth, they should see less of it up in the atmosphere. They suspected that by about 300m into the air, the extra radiation should drop off completely.

To a young, adventurous physicist this was the perfect challenge. All they needed was an instrument to detect radiation, and altitude. There was only one way of reaching high altitudes in the early 1900s if you weren't a mountaineer, and that was by hot-air balloon. At least three different researchers quickly took to the skies hunting for the background radiation, taking with them simple electroscopes for their studies,[1] but they were all unsuccessful. The balloon's motion shook the electroscopes and the pressure change caused them to spring air leaks and develop electrical isolation problems.

Electroscopes were popular because they could be made cheaply by almost anyone. All it took was a metal rod mounted inside a sealed container like a jar so that it was electrically insulated. Two fine pieces of gold leaf would be hung from the end of the rod. When a charged object – like a glass rod rubbed with fur – touched the top, the charge transferred down onto the gold leaves, causing them to splay apart from electrical repulsion and form an upside-down V shape. The leaves would stay that way for ever if the device was perfectly insulated. To measure radiation you simply charged the electroscope, then brought a radioactive sample near it, which would ionise some of the air inside and cause the leaves to lose their charge and slowly drop back down towards each other. The rate the leaves dropped at translated to the amount of radiation the device was exposed to. They were quite clearly designed for a stable lab bench, not to be taken up to altitude in a balloon.

Following this lack of success and growing confusion, German Jesuit priest and physicist Theodor Wulf realised the solution lay in making the electroscope more robust. In 1909, Wulf altered the electroscope to use two thin platinum-coated wires instead of gold foil. This proved much more reliable. Wulf travelled to Paris to test out his instrument at two different altitudes. First, he stood at the base of the Eiffel Tower and measured the radiation level there. Then, he ascended the tower and at 300m elevation, just where they expected the radiation to cease, he found the radiation persisted. Others adopted his method, but their results were just as perplexing. Italian physicist Domenico Pacini decided to go downwards in altitude, rather than upwards, and took a Wulf electroscope underwater, where he expected to find more radiation once he was surrounded by the minerals of the Earth. He found the opposite. The improved electroscope seemed to be working, yet the results weren't what they expected. A few scientists started to entertain the idea that the radiation wasn't coming from the minerals in the Earth after all.

Among them was twenty-nine-year-old Austrian physicist Victor Hess, who now saw his opportunity. He hired a hot-air balloon pilot, wrapped himself in a woollen coat, and launched from a field outside Vienna. The balloon soared to an altitude of over 5,300m, well above the height of Everest base camp. To his hot-air balloon, Hess had strapped two new Wulf electroscopes, specially adapted to handle the temperature and pressure changes. Despite the thin air and the temperature of around -20 degrees Celsius, he managed to take regular measurements, and eventually descended.

Hess wasn't the first to go to such altitudes or to try to measure the radiation levels in the atmosphere, but he was the first person to do it well enough to get a reliable result. Back on the ground, Hess could look at what he'd recorded. As he ascended, the amount of radiation had dropped off a little at first, but then it started to grow and grow until it became clear that there was far more radiation at high altitudes than at low ones. The radiation couldn't have been coming from the Earth – it must have been coming from outside the atmosphere. But from where? Hess made another balloon ascent during a solar eclipse which ruled out the Sun as a possible source. He was measuring a brand-new source of radiation. Now Hess, Wulf, Pacini and other physicists realised that radiation wasn't just to be found in minerals, or in the laboratory. There was radiation coming from space.

The radiation that Hess found, called *cosmic rays*,[2] solved the mystery of the excess radiation that had plagued physicists for more than fifteen years, but in doing so changed their view entirely of where radiation could be found. When I say radiation in this context, I should be clear that what I mean is *ionising* radiation, that which has enough energy to liberate electrons from atoms. This includes the three types of radiation scientists knew of at this point: alpha radiation (helium nuclei), beta radiation (electrons) or gamma radiation (high-energy light). Somewhere out in space, violent and forceful interactions were throwing off radiation powerful enough to travel across vast distances, through the atmosphere and down to Earth. But from where? How was the radiation formed? Was it a new type or one they already knew? Did it interact in the atmosphere, or travel straight through it? Hess had found cosmic rays, but he had little to say about their nature. What was needed was an instrument to learn more about radiation, both from cosmic rays and in labs on Earth.

What Hess and his colleagues really desired was some way to *see* radiation, which was particularly challenging because it is, for the most part, invisible. Yet they knew that physics had made visible other parts of nature through the use of clever instruments. For instance, the depths of space could not be seen before the telescope allowed the faint light to be collected, widening our view of the Universe and our place in it. The workings of biology were also invisible until the first microscopes emerged, bringing into view a teeming world of microorganisms, which led to enormous leaps in understanding the transmission of disease and the formation of life itself. Now, in the early 1900s physicists found themselves at a similar precipice, in need of a breakthrough in their ability to visualise radiation.

Charles 'C. T. R.' Wilson was a shy, diminutive Scottish physicist who began his scientific career around the time radiation was discovered. His heritage played an important role in the development of his ideas, primarily because Scotland happens to be an almost perfect place to study clouds. In 1894, aged twenty-five, Wilson travelled to Fort William to the highest mountain in the British Isles, Ben Nevis.

For 355 days a year the distinct flattened top of Ben Nevis is shrouded in treacherous mist, but Wilson arrived to something of a miracle: it was an unprecedented period of fine weather. He successfully hiked up Ben Nevis and stayed there for two weeks to volunteer at the meteorological station. Despite working at the Cavendish Laboratory in Cambridge, his first love wasn't physics, but meteorology. From the summit the clouds were mostly beneath him and from this vantage point he watched the effects of light dancing on the clouds, observing coloured rings called 'glories' forming in the shadows of the mountain on which he stood. He was fascinated by these effects and wanted to reproduce and study them in the laboratory. His first task, therefore, was to figure out how to make clouds.

Back in Cambridge he built an experiment to do just that. It was made with an upturned beaker inside a large glass jar filled with water, and a series of glass pipes and valves connected to a second jar held under vacuum. To operate the chamber Wilson would pull a wire, dislodging a small cork which would allow the air in the beaker to expand, dropping the pressure and temperature suddenly.[3] Anyone who has opened a soda bottle and seen a fog develop in the top as it makes a fizzing noise will have seen the same process in action. As air expands with the drop in pressure, it becomes supersaturated. If the conditions are right, the moisture in the air condenses onto particles of dust and forms fine droplets, producing a cloud. Wilson successfully reproduced this in the lab and was about to move on to recreate the effects of light he'd seen from the top of Ben Nevis when he found something he hadn't been looking for: when he used dust-free air in his experiment, the cloud droplets still formed.

How was that possible? To form clouds there needed to be some disturbance to initiate the formation of droplets; technically speaking, there needed to be some kind of *condensation nuclei*. Up until now that had been dust. But with dust-free air, what was causing the droplets? From his earlier experiments Wilson was able to tell that whatever was causing the disturbance was small, perhaps the size of the molecules or atoms in the air, leading him to the tantalising idea that the cloud-forming droplets might be forming on ions within the chamber. If that were true, he might have found a way of making visible and counting the atoms or molecules in the air.

Wilson wasn't interested in observing radiation. It was 1895, and radiation was so new that it wasn't well understood. He pursued his hypothesis that ions in the air were responsible for cloud formation. He rebuilt his experiment with a more elaborate setup designed to cause an even faster expansion. When his new experiment was ready, Wilson grabbed a primitive X-ray tube and pointed it into the chamber. He found that with the right conditions the X-rays produced large numbers of droplets, amplifying the effect he'd seen before. The presence of electric charges was causing clouds to form. His hunch was confirmed: the X-rays were creating ions in the air, and these ions created condensation nuclei.

Wilson was working on this as other physicists took electroscopes up in hot-air balloons and tried to solve the cosmic radiation mystery. He wasn't ignorant of the developments in radiation – after all he must have seen Ernest Rutherford and J. J. Thomson daily. At one point in 1901 he got interested enough to take a detour to search for the background radiation with an electroscope, which he set up inside the Caledonian Railway tunnel at night-time. Like the others, he was looking for extra radiation from the minerals of the Earth, but he saw no noticeable difference in the amount of radiation in the tunnel from what he had found in his lab.[4] He shifted attention back to his more promising work, leaving others to grapple with the mysterious radiation.

Wilson's seeming lack of interest in radiation and his strange cloud-forming experiment made him something of an oddity in the Cavendish Laboratory. He spent his days doing careful and complex glassblowing that would often break. The students and staff could all empathise with him, as they all had to learn glassblowing in the so-called 'nursery lab', a special laboratory where research students would learn the intricate skills of building devices like electrometers before embarking on a programme of replication of experiments that had taken place before. Many of them would later recall fondly the background sound of Wilson blowing glass, which formed almost a soundtrack to their time working at the Cavendish.

A scientific glassblowing workshop is a rarity today, so it is hard for us to appreciate what making an experiment like a cloud chamber involved before the existence of the computer-aided design and milling machines that we use to make modern experiments. It took years to master the techniques required, but Wilson's characteristic patience and gentleness would lead to him to create what Rutherford called 'the most original and wonderful instrument in scientific history'.[5]

Making a glass component was an artisanal process which involved heating the glass to the right temperature. In one hand Wilson would have held a blowtorch. To create enough heat to get the glass to melt just precisely as he needed, he would open up the gas a little, making the torch emit an unmistakable whoosh: the sound he would later be remembered for. At precisely the right moment he would use his mouth to blow air through a pipe, expanding the glass vessel with just the right amount of force, while he coaxed the molten glass with knives and tools as he worked.[6]

It was a hot and physically demanding process, but in just a couple of minutes Wilson could artfully reshape glass into a flask, a spherical bulb or a coil. The main pieces of his cloud chambers were cylinders that had to fit perfectly together, which often required hours of laborious grinding of the glass after it was cooled. By far the most treacherous process was in joining the various pieces together, when each new piece risked destroying the whole. More often than not, the whole experiment would end up smashed on the floor. Wilson, unlike Rutherford, would never utter profanities at his apparatus. He'd just quietly say 'dear, dear', and begin again.

Today, Wilson's early cloud chambers are preserved in the museum in the New Cavendish Laboratory in Cambridge and they look rather primitive at first glance. Their simple nature gives the impression that these were the easy days of discovery, when any half-decent physicist could make a groundbreaking revelation about the Universe. But once we understand the level of skill and patience required to make anything useful made of glass in the early 1900s, Wilson and his fellow experimenters start to seem quite extraordinary. Using this powerful new instrument, discoveries emerged that changed our view of matter for ever.

When Wilson first developed the chamber, it was far from clear that the device could be used to study radiation in any serious quantitative way, even if it appeared to respond to X-rays. It was only after Rutherford had determined the nature of alpha and beta radiation that Wilson turned back to the cloud chamber in 1910, this time with a renewed energy and an ambitious goal. He planned to make it a useful instrument for seeing charged particles.

In 1911, fifteen years after he invented the cloud chamber, Wilson became the first person to see and photograph the motion of individual alpha and beta particles. He had perfected the device so the charged particles would now create white trails that could be illuminated and photographed. He described these trails, produced by electrons in the chamber, as 'little wisps and threads of clouds'.[7] Wilson showed a photograph of the tracks of alpha particles to Australian-British physicist W. H. Bragg, who had first predicted that an alpha particle should initially slow down gradually before coming to an abrupt halt, interacting most strongly at the end of its path, producing a cloud trail which increases in opacity and thickness as the particle decelerates and eventually stops. Wilson and Bragg found that 'the similarity between the photograph and Bragg's ideal picture was astonishing'.[8]

Researchers around the world adopted the cloud chamber slowly but steadily, making adaptations so that it became even more useful. By the late 1920s most cloud chambers were placed between the poles of a large magnet, causing the trails from the charged particles to curve. A positive particle would bend in one direction, a negative particle the opposite way, and a highly energetic particle would curve less than a low-energy one. With careful measurements of the photographs, researchers could determine the electric charge and energy of the particles. In the lab they learned what different particles looked like in a cloud chamber and could determine their properties.

Ideas about particle interactions that had been acquired through painful experimentation over the years could now be seen playing out in front of physicists' eyes. The time was ripe to apply this new technique to understand the nature of cosmic rays.

At Caltech in Pasadena, Robert Millikan – who had moved there from Chicago in 1921 after his experiments on the photoelectric effect (Chapter 3) – encouraged his former PhD student Carl Anderson to use a cloud chamber to follow up on some intriguing work on cosmic rays. In 1929, Russian scientist Dmitri Skobeltzyn had found some tracks in a cloud chamber which hardly bent at all,[9] which indicated they had an enormous energy of over 5,000MeV, exceeding lab-based radioactive sources by a factor of a thousand. They weren't just energetic: they also appeared in surprising groupings, with two, three or more rays seeming to come from a point just outside the chamber. His results suggested that the cloud chamber might be able to reveal something new and exciting about cosmic rays.

Anderson, the son of Swedish immigrants, had decided while a young schoolboy in Los Angeles that he wanted to be an electrical engineer, despite having no family background in technical fields. A teacher encouraged him to go to Caltech, where he realised that physics was much more than pulleys and levers. He decided to switch majors and never looked back.[10] Anderson had already used a cloud chamber during his graduate work with Millikan, and had discovered that using alcohol vapour instead of water vapour made the trails much brighter and easier to photograph. He began building a new cloud chamber to do the job.

Anderson found a motor-generator from the aeronautics department and designed the rest around it. There was no money for fancy engineering – it was the start of the Great Depression after all – so his experiment was a beast of a thing, but it worked. The cloud chamber lay at the heart of the device, surrounded by copper pipes which conducted electricity to create a big electromagnet. The pipes were hollow, with water flowing through them to keep the magnet from melting. Together with the iron-pole pieces needed to direct the magnetic field, the experiment was the size of a small car and weighed about 2 tons. The chamber itself was visible through a hole in one end of the magnet, where a camera could take photographs of the cloud trails. To operate it, Anderson had to repeatedly create a very quick expansion of the alcohol vapour inside, which he did with a movable piston, resulting in a loud bang every time the thing operated. The rest of the campus at Caltech would have heard the repetitive banging ringing out from the roof of the building, where it had been assembled. Thankfully for the rest of the occupants, Anderson only operated the experiment at night-time because it required 425kW of electricity to run – a substantial portion of the power consumption of the whole campus.

Trawling through his photographic evidence, Anderson found that around fifteen of the 1,300 photographs appeared to show tracks that corresponded to positively charged particles. But the tracks were too long for the lightest known positively charged particle, the proton. What was this seemingly new particle? The particles in his photographs had one unit of positive charge and a mass similar to that of the electron. At first he simply called them 'easily deflectable positives', but by the time he'd written up the publication, he had come to a surprisingly bold conclusion. He believed he had found an entirely new type of fundamental particle, which he called the 'positron'.[11]

What Anderson didn't know was that a couple of years earlier, in 1928, the British physicist Paul Dirac had predicted, purely through mathematical insight, that positrons should exist. Dirac combined the theory of quantum mechanics, to describe things that are very small, with Einstein's theory of special relativity, to describe things that travel very fast, in the hope of gaining insight into the atom. Dirac was a man of few words, but his work managed to unify two of the most talked-about new theories in physics. The equation that he produced, known simply as the Dirac equation, is considered by many to be the most beautiful equation in all of physics. It also had unintended predictive power. Just as the square root of four can have the solutions +2 or -2, the Dirac equation predicted that there should be particles that were identical to the electron – that is, having the same mass – but with opposite electric charge. Because of the strange implications, Dirac wasn't sure of the physical manifestations of his theory, but it seemed to predict that every known type of particle should have an opposite version, which became known as *antimatter*.[12]

Dirac happened to be friendly with one of the experimental physicists in the Cavendish Laboratory, Patrick Blackett, who was further developing the cloud chamber technique along with Italian-born physicist Giuseppe Occhialini. When Dirac came up with his new theory, he shared it with Blackett and together they had worked out that if a positron were to appear in the magnetic field of a cloud chamber, it would leave a track that looked identical to an electron but curve in the opposite direction. Almost three years before Anderson's work, they looked through Blackett's cloud chamber photographs from experiments with radioactive sources in the lab. Dirac thought there seemed to be plenty of evidence for positrons, but Blackett thought the evidence far too uncertain to publish. There might, he argued, be electrons coming from outside which by chance collided in such a way that they looked like positrons. There was no way for him to tell the difference between these wayward electrons and real positrons without re-doing the experiments.[13]

Blackett may also have been wary because Dirac's idea hadn't exactly been well received. Some of the scientific greats of the time were – to put it kindly – unconvinced about the idea that our Universe might consist of two types of matter, 'normal' matter and a mirror 'antimatter'. Austrian physicist Wolfgang Pauli, one of the pioneers of quantum theory, called it 'nonsense' and Niels Bohr (see Chapter 2) was 'completely incredulous'.[14] Werner Heisenberg, a German theoretical physicist who created much of quantum mechanics, including the uncertainty principle, stated in 1928 that 'the saddest chapter of modern physics is and remains the Dirac theory'.[15] Blackett went away to do some calculations to determine whether or not they really had evidence for Dirac's extraordinary theory, but while he was still thinking about the problem, the news reached them that Anderson had discovered the positron.

Anderson was too busy with his experiment to have read Dirac's papers. Perhaps this focus was well founded, given that he managed to beat Blackett and Occhialini to the positron discovery. However, his results were hotly debated by the physics community because a few individual photographs seemed scant evidence to back up such an extraordinary theory. With this in mind, the Cambridge team realised they had an advantage. Rather than collecting thousands of photographs with the hope that a handful of them might show something of interest, as Anderson did, Blackett and Occhialini had worked out how to get about an 80 per cent success rate at capturing interesting particles zipping through the chamber. To do so they had developed an electrical method to 'trigger' the camera, placing a Geiger counter above and below the device so that if both counters detected a particle around the same time, the chamber would be photographed. By 1932 they had both the method and the imperative to follow up on Anderson's work with experiments of their own.

Blackett and Occhialini quickly confirmed the existence of positrons, and with their data-rich observations they were also able to delve deeper into the details. What they observed were many events where electrons and positrons were found in photographs together. In fact, there seemed to be equal numbers of electrons and positrons in the photographs: normal matter and antimatter were created in equal quantities. Blackett and Occhialini were able to observe this process in action when high-energy gamma rays (present in cosmic rays) entered the chamber and created an electron and a positron at the same time in a process known as *pair production*. This was the first observation of photons (gamma rays) transforming into matter (electrons and positrons), a process predicted by the combination of quantum mechanics and Einstein's relativity. The existence of these interactions revealed a second dazzling consequence of Dirac's equation that was only just dawning on theorists at the time: antimatter and matter can *annihilate* each other when they come into contact, turning their mass into energy, emitted as light. In other words, mass can be converted into energy and vice versa. They collected so many photographs of positrons and of pair production that the scientific world could no longer resist the conclusions from Dirac's theory. Strange as it seemed, antimatter was real.

Rather than rewrite the story and ascribe to himself some kind of insight after the fact, Anderson insisted that 'the discovery of the positron was wholly accidental'.[16] It was one of those discoveries that was ripe for the taking and would have happened somewhere else soon had he not made it first. Alongside Victor Hess, Anderson collected a Nobel Prize in 1936 at the age of thirty-one, the youngest person ever to win a physics Nobel. Charles Wilson had been awarded one in 1927 for the invention of the cloud chamber and Dirac received one in 1933.[17]

With his first attempt, Anderson had made a remarkable advancement using the cloud chamber to explore cosmic rays. But this was not the end of the road. The discovery of the positron hinted that the investigation of cosmic rays would lead in interesting directions – that cosmic rays could be used to discover hitherto unknown particles, and that nature was richer and more abundant than they had previously known.

The positron experiment had shown what could be detected at ground level, but still little was known about cosmic rays themselves. Anderson set off on a new adventure in 1935 with the cloud chamber, this time with his own graduate student, Seth Neddermeyer. To study high-altitude cosmic rays, Anderson and Neddermeyer went to Pike's Peak in Colorado.

Their plan involved working at 4,300 metres elevation, where the oxygen level is just 60 per cent of that at sea level, exposing themselves to a high risk of altitude sickness. The climate on Pike's Peak is also inhospitable. It's snowy for almost the whole year and when winds hit, which they frequently do, they can blow at up to 160kph (100mph). To make matters worse, Anderson and Neddermeyer still had almost no money.

They managed to scrape together enough to buy a flatbed truck for $400, mounted the huge experiment on the back of it, and set off across the country to Pike's Peak. All went smoothly until they started climbing the mountain itself. Weighed down and with low oxygen levels at altitude, the old truck couldn't make it up the mountain. They had to be rescued and towed. When they eventually reached the top, they discovered there wasn't enough electricity to power their experiments, so they bought another car and used its engine as a generator.

When they were finally up and running, the two physicists captured images for six long weeks. They then had to develop photographs from negatives before they could get any hint of what they were capturing. On the cold and dark mountain they studied the images to look for electrons, positrons, protons and alpha particles. This time, they kept finding tracks of particles that looked a lot like electrons, but seemed to be about 400 times heavier and had both positive and negatively charged versions. They knew these particles weren't protons – they were too light for that – and they weren't the newly discovered positrons either. There was only one conclusion they could draw: they had discovered another new type of particle.

We now call these particles 'muons'. They have exactly the same properties as electrons (or positrons for the anti-muon) but are heavier in mass. They also have a limited lifetime, decaying in 2.2 millionths of a second and transforming into electrons.[18] When high-energy cosmic rays hit the atmosphere their collisions create showers of new particles, vast numbers of which are muons. Every minute of every day around 10,000 muons bombard each square metre of the surface of the Earth (a few go through your head every minute) and yet we can't see, feel or otherwise detect them without specialised equipment. There are even more of them at altitude than down on the ground.

Unlike electrons, protons and other particles that had been observed, there seemed to be no apparent practical reason for the existence of muons. They are a fundamental particle; that is, they aren't composed of other particles, but they don't make up any part of the normal matter around us. On finding out about the muon, one physicist at the time responded 'who ordered that?'[19] Their reason for being was, and still is, a complete mystery. There was a depth and complexity to the subatomic world that physicists were only just starting to come to grips with.

One idea of what a muon was illustrates where the theoretical understanding was in 1935. A young Japanese theorist named Hideki Yukawa had proposed the previous year that the force that holds the nucleus together – the strong nuclear force – was due to a particle with a mass roughly 200 times that of the electron. This proposed particle he called the *meson* from the Greek word for 'intermediate' because he predicted it should have a mass somewhere between the electron and the proton.[20] At first, some physicists thought the muon was the Yukawa meson, but they soon realised it couldn't be, as the meson should interact strongly with matter. The muon, on the other hand, appeared to travel straight through sheets of lead and other materials.

Getting the best vantage point and the best data meant highly adventurous experiments which pushed the boundaries of technology. Anderson's cloud chamber experiment was later even mounted and flown in a Navy B-29 to study cosmic rays at high altitude,[21] although the engineering issues were so great that the experiment didn't produce many usable results. Over time it became clear that the particles that make up the matter of our everyday existence are only part of the hidden world around us. There is much, much more that lies beyond. Where the discovery of radiation had changed the view of matter from a static substance to one of constant change, now cosmic rays began to shatter the idea that atoms were the only type of matter that existed. The muon was just the start.

Getting high up in altitude to detect cosmic rays before they interacted with Earth's atmosphere became increasingly important as knowledge of cosmic rays and new particles grew. As the B-29 experiments showed, a more robust type of detector than a cloud chamber was needed for high-altitude experiments. Other physicists were hard at work developing another complementary type of detector. Unlike the complex pistons and cameras involved in cloud chambers, these *nuclear emulsions* were passive detectors with no moving parts. They were essentially a special type of photographic plate, with silver halide crystals suspended in gelatin, which are sensitive to the passage of charged particles. Nuclear emulsions were more robust than cloud chambers and far less onerous to work with – they could be left unattended accumulating data for months at a time, and could even be launched high into the atmosphere without concern.

The method of using these emulsions to study cosmic rays was developed by Austrian physicist Marietta Blau while she worked, unpaid, at the prestigious Radium Institute in Vienna. She had completed her PhD in Vienna in 1919 working under Franz Exner and Stephan Meyer, both of whom had a reputation for supporting women scientists.[22] She had started out on a promising career at Frankfurt University, teaching radiology to medical students and publishing research on photographic emulsions for X-rays and visible light. After she returned to Vienna in 1923 to care for her ill mother, she took on her unpaid role at the Institute since she could find nothing else, and supported herself with grants and bits of college teaching.

Blau's research in Vienna combined what she had learned in Frankfurt with her knowledge of the emerging field of nuclear science and showed that photographic emulsions could be used to study cosmic rays. She worked with emulsion manufacturer Ilford to produce extremely thick versions which were better for recording particle tracks. With her former student Hertha Wembacher she ran an experiment for four months at the Hafelekar research station in the Austrian Alps. The results showed a remarkable new discovery of 'stars of disintegration', left behind when cosmic rays collided with heavy nuclei within the emulsion and caused them to explode into star-shaped arrays of particle tracks.

Sadly, her work was very soon cut short. Blau was Jewish and on the eve of the Anschluss in 1938, she fled from Austria and went to stay with chemistry pioneer Ellen Gladisch in Oslo, later moving to Mexico and then to the United States with Einstein's assistance. Meanwhile, her collaborator Wembacher, a Nazi party member, continued publishing their results but removed all reference to Blau.

On the other side of the world, Blau's technique was taken up by another woman, Bibha Chowdhuri, an Indian researcher who completed a Masters degree in physics in 1934. This was still a rare achievement for a woman anywhere in the world, including in India. When Chowdhuri first asked to join the research group of D. M. Bose, she was told that he had no projects suitable for women. She persisted and, from 1939 to 1942, Chowdhuri and Bose conducted cosmic ray studies by leaving photographic emulsions for months at a time at high altitudes in Darjeeling, Sandakhpu and elsewhere. The emulsions had to be painstakingly developed and processed, which could take months of work with a microscope. Chowdhuri and Bose found evidence for two new subatomic particles with masses roughly 200 and 300 times the mass of the electron. We have already seen the first of these, the muon, but the second one is new in our story: the pion. There are three types of pions (positive, negative and neutral) which we will return to in more detail in later chapters as we unpick the new particles and the forces through which they interact.

Despite being the first to recognise the particle, Chowdhuri's contributions were not acknowledged by the scientific community. In 1947 British physicist Cecil Powell (with Giuseppe Occhialini) used the same method, albeit with superior emulsions, to demonstrate the existence of the pion. In 1950 the Nobel Prize committee awarded Powell the physics prize for 'for his development of the photographic method of studying nuclear processes and his discoveries regarding mesons made with this method',[23] but Chowdhuri did not receive a Nobel nomination. The reason that Chowdhuri's experiment was not taken by the Nobel Committee as the key discovery of the pion appears to be because the quality of the emulsions she used was not quite sufficient for completely unambiguous discovery, a fact caused by supply issues in the Second World War.[24] But it takes only a quick search to find Powell's reference to her work in one of his key papers[25] and acknowledgement of the precedence of her work in his book on elementary particles.[26]

Blau was nominated for the physics Nobel a few times for her invention of the photographic emulsion technique, which the community – including Powell – recognised as essential to their progress in understanding high-altitude cosmic rays. Her invention, produced in large quantities by Ilford and Kodak, led to widespread use of photographic emulsions and was essential to Powell's pion discovery. Yet biased reports of her contributions in an overtly negative evaluation by a member of the Nobel committee[27] meant that Blau was overlooked too.

Blau, Chowdhuri and others are not anomalies. There have been so many incidents of the work of women in science being unrecognised or ignored throughout history that this effect even has its own name: the Matilda Effect. It was coined in 1993 by historian Margaret Rossiter[28] after American suffragist Matilda J. Gage, who first articulated this phenomenon in the late nineteenth century about female innovators. Rossiter hoped that by giving the effect a name, it might encourage historians, sociologists and – one would hope – scientists themselves to include more stories of systematically forgotten women in science, or perhaps work to bring more of them to light.

Teams of physicists around the world continued to study cosmic rays using cloud chambers and photographic emulsions over the next two decades, slowly teasing out their properties. Cosmic rays are known to have extra-terrestrial origins and yet even now, almost a century later, their formation is poorly understood. Information from the Fermi Space Telescope gives evidence that they may be formed in supernovae, and may be boosted to high energies in the gravitational fields around black holes. However they are formed, we know that they mostly consist of very high-energy protons. These protons shoot down through the Earth's atmosphere and collide with atoms in the air, producing an avalanche of other particles: the muons and positrons are these 'secondary' particles. Almost all the protons and many of the secondary particles either interact with the air or decay (muons have a lifetime of 2.2 microseconds[29]) before reaching the ground, which is why the early pioneers detected fewer cosmic rays at ground level.

Cosmic rays carry enormous amounts of energy, so much so that they readily smash apart atoms. If that happens in the right location, as Marietta Blau first discerned, scientists can watch the resulting pieces of these collisions and learn about the nature of the atom and other particles. We now know that many cosmic rays have travelled light years across the Universe, bringing with them information about what is out there in astronomical systems like neutron stars, supernovae, quasars and black holes.

Here on Earth, we are completely oblivious to the shower of cosmic rays, and yet around 100 of them pass through our bodies every second.

A billion billion cosmic rays strike the Earth every second, a power of over a billion watts. If you were somehow able to harness that power and added it up in terms of kW (kilowatts – a washing machine runs at about 1kW over an hour-long cycle), it combines to 3.6 billion kW every hour, or around 32,000 TWh (Terrawatt hours) each year – about 50 per cent higher than the electrical power consumption of the entire planet in 2018.

As different particles and forces have been discovered, one thing has remained true: the scientists who discover them almost invariably believe they will have no practical use. Just as J. J. Thomson couldn't see a use for an electron, it has taken a long time to figure out the value of cosmic rays. Now, more than a century after first finding cosmic rays and almost 80 years after the discovery of the muon, advances in technology have led to an understanding of how cosmic rays interact with the Earth and to real-world applications of both positrons and muons.

Cosmic rays can tell us more about the history of life on Earth. The effect of cosmic rays on nitrogen in the atmosphere creates a radioactive isotope of carbon, called carbon-14. This combines with oxygen to form carbon dioxide, which plants absorb through photosynthesis. Animals and people then ingest these plants, absorbing mostly regular carbon-12 but along with it, a small amount of carbon-14. In the 1940s Willard Libby realised that by comparing the amount of carbon-14 to carbon-12 in a sample of wood, bone or other organic material, it was possible to calculate how long ago an animal or plant had died, as the carbon-14 would decay with a half-life of 5,730 years. Radiocarbon dating, which we'll discuss more in the next chapter, has had a profound impact on archaeology by making it possible to create a global timeline on which events in different regions and continents can be placed. As a result, we now have a prehistory not just of individual regions, but of the whole world.

The interaction of cosmic rays can also tell us about the history of the Earth's climate and its changes throughout geological time, in particular to tease out the influence of the Sun. The Sun isn't the source of high-energy cosmic rays – we've known that for over a century, since Victor Hess flew his balloon during a solar eclipse – but the Sun does influence how many cosmic rays arrive at Earth. We now know that the Sun is constantly throwing off material, called the solar wind, creating the *heliosphere*, a vast bubble in space surrounding the planets in the solar system. When the Sun's activity is quiet the heliosphere is weaker, and it lets more cosmic rays into the solar system, which go on to collide with atoms in Earth's atmosphere.

When cosmic ray protons hit oxygen in the atmosphere they can create two isotopes of beryllium: beryllium-7 and beryllium-10, which end up deposited on the Earth. Beryllium-10 has a half-life of 1.4 million years and decays to boron-10, and beryllium-7 decays in just 53 days to lithium-7. These isotopes build up in layers of ice in Antarctica and Greenland, and drilling down into the ice to take cores supplies a neat method to trace these back in time. For each layer, the ratio between the two isotopes determines how long ago they were formed in the atmosphere, and the amount of beryllium-10 tells us how active the heliosphere, and thus the Sun, has been. Using this method, cosmic rays can tell us if solar activity is indeed linked to climate change on Earth.

Transformational everyday uses have also been found for the particles discovered from cosmic ray studies. Positrons, which are emitted naturally in some radioactive decay processes, are used to detect and understand disease through the technique of Positron Emission Tomography (PET). Machines that conduct these detailed medical scans can be found in most major hospitals, and we'll also learn more about this application in a later chapter.

By far the most unexpected particle to find an application is the muon. Muons have a unique feature, in that they can travel a long way through dense objects – a lead wall or a few hundred metres of rock are no hindrance to them. As technology developed, physicists realised that if they could set up detectors in the right way, they could use the muons from cosmic rays like a huge X-ray scanner. Because they can travel all the way through massive objects, they can do things that X-rays just can't do.

Muons were first used not in the United States or Europe, where they were discovered and studied, but, somewhat surprisingly, in Australia. In the 1950s a physicist named E. P. George used cosmic ray muons to measure the density of rock above a new tunnel for the country's enormous hydroelectric power system, the Snowy Hydro scheme. With a Geiger counter he was able to detect cosmic ray muons in the tunnel and on the surface, and then used the result to measure the depth and density of the ground between. But the Geiger counter he used gave no information about the direction the muons came from, so it wasn't possible to make any kind of image.

By the 1960s Luis Alvarez (who was also the founding research director of CERN, which will come into our story later) collaborated with archaeologists to use muons to image the insides of pyramids, which eventually led in 2010 to the 'ScanPyramids' project by Cairo University and the French Heritage Innovation Preservation (HIP) Institute. While archaeologists thought they had learned everything there was to know about Khufu's Great Pyramid in Giza, in 2017 the ScanPyramids team placed muon detectors around the pyramid, as well as inside the Queen's Chamber within, and came to a surprising conclusion: there is a hidden room, disconnected from all others, inside the structure. It was the first new room to be discovered since the nineteenth century.[30] Their findings constitute a breakthrough in understanding the internal structure of the pyramid and a possible step towards finally understanding its construction.

Compared to electrons or X-rays, muons don't interact very much as they travel through matter, so they are less prone to scatter and mostly travel through objects in straight lines. This lack of scatter gives them a surprising advantage. Placing detectors on either side of an object and correlating the passage of a muon before and after they travel through the object produces images with surprisingly good resolution, even if there aren't many of them. This is because of the certainty that they have travelled in straight lines, compared to X-rays which will always have much more scattered trajectories. The first images made this way came from developments in the United States, and new and better detection devices gave us the ability to see inside large closed volumes with a technique known as muon tomography or 'muography', which works like a 3D X-ray scanner but on an enormous scale. In the 2000s, research and application in this area increased dramatically.

In 2006 a Japanese team led by Professor Hiroyuki Tanaka from the University of Tokyo became the first to use muons to take an image of the internal structure of a volcano, Mount Asama in Japan. Geologists have been particularly strong adopters of muography, and other volcanoes including Etna and Vesuvius have been imaged to map out lava channels and predict eruptions. They can now conduct time-sensitive imaging, allowing them to see the magma moving.

As the technology has matured, muography has been commercialised, often with the formation of spin-off companies from laboratories where the research was done. These companies have found vast and fascinating applications of imaging using muons, producing 3D visualisations of everything from whole container ships to critical infrastructure like power stations.

Muon detection systems are also on the market for national security agencies and for mining applications, identifying dense mineral deposits, caves and tunnels, or other structures in the Earth. Muons can be used for geophysics, groundwater mapping and minerals exploration. In nuclear safety, one of the first groups flown in after the 2010 tsunami in Japan was a team who used muography to analyse the state of the Fukushima Daichi reactor cores and locate the nuclear fuel, meaning fewer surprises in handling the clean-up and management of the incident. No other technique can make such an image. Other groups are looking at using the same approach to inspect nuclear-waste storage facilities.

We're only just starting to see the full benefit of the muons that travel unseen towards the Earth every day. In the future, muons may be used to monitor everything from the structural integrity of bridges to the rumblings of the Earth.[31]

While physicists today no longer use cloud chambers, these detectors kickstarted a remarkable investigation into the nature of cosmic rays and enabled the discovery of a range of new particles. The cloud chamber started out as a curious device aiming to recreate the effects of light on clouds, and ended up being the tool that physicists needed to view the invisible world of particles. For the first time physicists could *see* as particles traversed across their detectors, and could capture photos that showed particles coming into and out of existence.

Before the cloud chamber, physicists thought that the only particles were *subatomic* – inside the atom – but now they knew there were particles that play no role in the matter we find around us. The challenge ahead was to try to figure out if there were even more particles in nature, and how all these pieces fit together.

The biggest problem was that physicists still had no control over what they were observing. They were reliant on natural sources of particles for all their experiments, from radioactive substances to cosmic ray muons. But to delve further into the atom and understand the new particles found in cosmic rays, they would need to develop techniques to manipulate matter on the smallest scale. They needed to mimic cosmic rays in the lab.

5

The First Particle Accelerators: Splitting the Atom

Charles Bennett had picked up the violin for eighty dollars at a flea market in Rochester, NY. When he peered inside the intricately carved f-shaped hole on the front, he saw a distinctive yellow label that said 'Stradivarius'. Stories abounded of flea market finds like this: cheap antiques that turned out to be worth hundreds of thousands of dollars. Strangely, Bennett didn't think to get the violin professionally valued, but we can assume that all the relevant experts were in Europe and the shipping was more than a penniless graduate student could afford in 1977. It didn't take long before he realised that, in order to know the violin's true value, he would have to destroy it. Bennett was stuck. He might not have been a virtuoso, but he wasn't willing to render the instrument worthless. He went back to his physics PhD work.

To know if it was a real Stradivarius, he would have to verify the instrument's age, and Bennett knew about cosmic rays and the technique of carbon dating from his physics training. Stradivari usually used a combination of spruce, willow and maple. If Bennett assumed the trees were cut down not long before the instrument was made, he could use carbon dating to compare the amount of stable carbon-12 to radioactive carbon-14 left in the wood, and check whether his violin was really an early eighteenth-century masterpiece. Discussing it with his PhD supervisor, Harry Gove at Rochester University, they calculated that for every thousand billion carbon-12 atoms there would be only one carbon-14 atom. A sample containing a gram of carbon would only decay and give off an electron to count roughly once every 5 seconds. They considered chipping off tiny slivers from the violin, preserving the instrument and trying to measure something, but the count rate would be far too low. To make the method work, they'd have to cut out a huge chunk of the wood.

A few weeks later colleagues – who had no idea of the violin conundrum – came to visit Gove with the idea of using his nuclear physics lab for an experiment to measure the amount of carbon-14 in a small sample. The two colleagues – Albert Litherland and Ken Purser – had both worked with Gove on nuclear physics experiments in their earlier days, and both had independently come up with the idea of using an accelerator for carbon dating. A discussion with Gove at a conference a month earlier had spurred their visit: Gove had the experimental equipment and know-how to make their ideas work. The Rochester University particle accelerator, which dwarfed every other piece of equipment, could take small samples of material and create beams of the constituent atoms. Gove had never tried separating out the different carbon isotopes before, but if the proposed experiment worked he realised it might also open a path to putting a date to the violin without destroying it. Gove agreed to the experiment on the condition that Bennett would be part of the collaborating team.

To find out whether Bennett would indeed make his fortune, we need to understand how the particle accelerator they were using works. So far, all the experiments we have seen have been done with fairly simple equipment and radioactive substances found in nature. In this chapter, we'll begin to explore why elephant-sized equipment was suddenly required to understand the smallest constituents of nature. By the mid-1970s, when Bennett and Gove came upon the dilemma of the violin, these machines had been the workhorses of nuclear physicists for decades and were even being used in many areas of science and industry in completely unforeseen ways. But that all came many years after they were first invented. Back at the Cavendish Laboratory in Cambridge in the 1920s, the journey towards particle accelerators began with one of the biggest questions about the nature of matter: what is inside the atomic nucleus?

Rutherford had taken over at the Cavendish from J. J. Thomson in 1919, and the general feeling in his lab since then had been one of experimentation as usual. But beneath the surface was a current of frustration. Back in 1911 Rutherford had described the existence of the nucleus and then devoted himself to understanding this new heart of matter, expecting to make quick wins. Rutherford had become rather accustomed to seeing his name in news headlines on a regular basis as he made breakthrough after breakthrough. But now he hadn't had a big discovery for almost a decade.

The experiment by Geiger and Marsden which had discovered the atomic nucleus had made Rutherford the world expert in the subject of atoms. By the early 1920s Rutherford could put his knowledge together with that of the chemists, who had distinguished ninety different types of atoms, with some difficulty, based on their masses. Over time it emerged that the atomic masses of all the elements seemed to be direct multiples of the weight of the lightest element, hydrogen. Helium was four times heavier, lithium six times, carbon twelve times and oxygen sixteen times heavier. That couldn't be a coincidence. What's more, all that mass wasn't coming from the electrons, tiny and lightweight as they are. The *nucleus* must be the key to understanding the true nature of matter. The pattern of masses hinted that the nucleus itself might be composed of building blocks.

The one thing Rutherford knew for sure was that there were protons inside it. During the First World War, he'd conducted an experiment in which he fired alpha particles onto nitrogen gas, producing hydrogen nuclei. In 1917 he managed to show that all atoms seemed to contain hydrogen nuclei – positively charged particles which were thereafter called *protons*. The problem was that the atomic nucleus for elements heavier than hydrogen couldn't be composed of *only* protons. The positively charged protons would all push apart from each other, so the question was: what holds the nucleus together under such a 'repulsive' force? It was possible, Rutherford thought, that a neutral particle might somehow hold it together. As a result, an atom like helium with four times the hydrogen mass but a maximum electric charge of two (after knocking off its two electrons) might not contain four protons as he'd assumed. It might be that there are just two protons and two of a hitherto unknown particle as heavy as the proton, but with no electric charge. They dubbed it the neutron. Rutherford and his team searched for the neutron, but years went by and they found nothing.

Imagine an ego the size of Rutherford's at this point. He was a New Zealand farm boy who had won the Nobel Prize in 1908, had been knighted in 1914 and was now director of the world's pre-eminent physics laboratory. It was almost a matter of pride that he should be the one to answer the next big question. At this point cosmic rays had been discovered, but the muon and positron had not. Rutherford's focus was squarely on the atomic nucleus and he felt that there was only one way to make progress: they had to break apart the nucleus to find out what was inside. He didn't want to just chip protons off. He wanted to split the whole thing open.

The tools at Rutherford's disposal were the same as always: a source of alpha particles, a target and a scintillation screen. The particles were emitted from a source of radium or polonium, which would be sealed in a metal tube with a slit at the end to fashion a kind of alpha particle gun. While this provided control over the particle direction, most of the alphas would crash into the tube walls and be lost, leaving precious few to work with.

The industrious Cavendish students and researchers nevertheless pushed on with a relentless stream of experiments, hoping that the nucleus would give up its secrets. They would send the alphas carefully through various gases, ping them into metal foils and plates and bombard them onto just about anything they could get their hands on with the hope that they would see a reaction. A few light elements displayed the same result as nitrogen – coughing up a few protons – but heavier elements produced no results whatsoever. No neutron was found, and nothing surprising or spectacular emerged. The nucleus remained an enigma.

The Cavendish experimenters were stuck with a few measly alpha particles. There was no way to control the experimental parameters, since alpha particles always emerged from the radioactive decay of the source with the same energy, and they hadn't yet figured out how to produce anything other than the alphas that emerged naturally. Compounding the issue, alpha particle sources like radium were weak and only got weaker as they decayed over time. Some preparations would decay away in as little as half an hour. As tools for exploring the nucleus of the atom, they were starting to seem feeble.

The only thing they *could* control was the error in their experiments. To ensure that they made the most of the unreliable particles, the group came up with an intricate method for making reliable observations. A typical experiment would need three researchers. Two would go and sit in a dark room to let their eyes adjust. The third person would finish preparing the apparatus, then close the shutters and curtains when all was ready, and the experiment would begin. The two researchers would take turns looking through a microscope pointed towards the scintillating screen, each watching for roughly a minute at a time, marking any flashes on a scrolling roll of paper. After about an hour of this to-and-fro their eyes would grow tired and a new team would take over. It was arduous but necessary work, an evolution of the same technique that Marsden and Geiger had used many years earlier.

All the new research students in the Cavendish would be trained and tested in particle counting when they arrived, guided by Rutherford's meticulous long-term colleague, James Chadwick. As well as conducting his own detailed research, which we'll return to later on, Chadwick oversaw student training in the 'nursery' laboratory. When they were ready, they would speak to Rutherford, who would recommend a direction for a research project.

The students would build their experiments from scratch. Just getting parts for experiments from the Cavendish stores was a challenge, meaning they had to be resourceful and determined. The workshop manager, a man named Lincoln, guarded closely the lab's resources. He would measure out pieces of wire rather than hand over a roll, and count out individual nuts and bolts. Legend has it that one student who was looking for a piece of steel pipe was offered a saw and directed to the bicycles in the courtyard. In reality, the miserliness came from the top, from Rutherford, who would far prefer to impress everyone with a cheap but ingenious experiment than have to continually justify expenditure or beg for money.

For all their ingenuity, careful methods and resilience, they still continued to find nothing. One solution was for the Cavendish to acquire more radium. But the precious material was in short supply and Rutherford's thriftiness overruled the idea. The team knew their competitors had the advantage of having much more radium to work with. Marie Curie had been gifted an enormous supply of a whole gram of radium by the women of the United States as a tribute to her tour there, and her physicist daughter Irène Curie and Frédéric Joliot were hard at work using it in Paris. Numerous other laboratories in Europe were also trying to get ahead, but the Cambridge team had continued to maintain their leading edge through sheer hard work. That was, until 1924, when their status as the world's premier laboratory came into question.

A research group in Vienna circulated a paper that seemed to show that disintegrating atoms was easy. They were trying exactly the same experiments as the Cavendish team and producing astonishingly different results. Spirits took a nosedive in Cambridge. Under Chadwick's guidance all the student counters were re-trained and re-tested before they doubled down to try to reproduce the Vienna team's results. But they simply couldn't. A respectful but forceful disagreement between the two groups emerged and eventually Chadwick travelled to Vienna to get to the bottom of it.

In Vienna the researchers hired women specifically for the task of counting, but, unlike in Cambridge, they would tell the counters roughly what they were looking for before the experiment began. Lo and behold, the counters would find it. Repeating the experiments without this intervention, the Vienna group failed to replicate their earlier results, and their data fell in line with the Cambridge results.

With this episode behind them, Rutherford and Chadwick had to acknowledge a truth that was slowly becoming clear: their reliance on weak alpha sources was holding back their scientific progress. They knew there were discoveries to be had. To get there, the experiments needed a radical change. They needed a way to produce protons, alphas and other particles at will, with varying energies. But no such method yet existed. They would have to invent the technology themselves.

Ernest Walton had finished his training, and it was now time for him to take on a research project of his own. Walton was the son of a cler-gyman, a twenty-four-year-old Irish physicist who had recently arrived in Cambridge for his PhD, aided by the same scholarship scheme[1] that had once helped Rutherford move from New Zealand to the UK. Walton excelled at both mathematics and physics and had obtained a first-class degree in both subjects in Dublin. As he also liked to build things, experimental physics seemed the perfect fit. Plucking up his courage, he pitched his idea to Rutherford. He wanted to try to build a machine to accelerate charged particles.

Little did Walton know that two days earlier, Rutherford had been in London delivering a rousing address to the Royal Society in his new role as President. He stood before the esteemed gathering and declared the most important and pressing need for science as it stood that year, in 1927. He wanted to find a way to create 'a copious supply of atoms and electrons which have an individual energy far transcending that of the alpha and beta particles'.[2] If such a thing could be done, a beam current of just a milliampere could produce more particles than 100kg of radium could, an astonishing quantity. What he needed was a way to extract fundamental particles and launch them with high energy towards an atom. What he wanted was exactly what Walton had just pitched to him: a *particle accelerator*. Impressed by the young Irishman's courage, Rutherford agreed and immediately walked him downstairs to a laboratory where he could find some space to work.

The laboratory he'd chosen was a brick-walled space in the basement with a high ceiling. It contained three workbenches and already housed two other researchers, Thomas Allibone and John Cockcroft. Allibone, or 'Bones', had recently made a similar pitch to Rutherford and was already trying to use high-voltage Tesla coils to accelerate electrons to high speeds. Rutherford must have figured some friendly competition would help both young researchers along.

The other researcher was John Cockcroft, who, at thirty, was a few years older than the others because he had taken a rather circuitous route to the Cavendish Laboratory. Cockcroft was known for his ability to get things done, and his colleagues would regularly comment on the way he comfortably managed the workload of about two and a half full-time jobs. He was conducting his own research while also helping to manage the enormous experiments in the neighbouring lab of Peter Kapitza, who was attempting to create extremely powerful magnetic fields. Torn between different tasks, he would scribble reminders to himself in his little black pocket book in tiny illegible script. It was said by his colleagues that anything he wrote in there 'would be attended to promptly'.[3] He had been aware of the challenge to build a high-voltage device to accelerate particles, but now, after Rutherford's speech, the idea was firmly lodged in his mind and his notebook. He knew there were two huge barriers that would need to be overcome. One was theoretical, and the other was experimental.

Cockcroft was uniquely placed to solve a problem with both theoretical and practical challenges. The First World War had interrupted his studies in mathematics and afterwards he had taken an apprenticeship with a company called 'Metropolitan Vickers', or Metrovick, a heavy electrical engineering firm in Manchester who dealt in industrial equipment like generators, turbines, transformers and electronics. Only after this period moonlighting as an engineer did Cockcroft attend Cambridge University to finish getting a good grounding in mathematics and physics, and he emerged as both an experimental physicist and quite a respectable theorist.

The main theoretical problem they faced was how to fire alpha particles or protons into the atomic nucleus, when these positively charged projectiles would be repelled electrically by the positively charged atomic nucleus. This repulsion is known as the *Coulomb Barrier*. Cockcroft first had to calculate the energy the alpha particles would need to break through this barrier and gain entry into the nucleus. He knew from his theoretical work that this figure would translate directly to the amount of voltage required to accelerate alpha particles with sufficient energy. He did the calculation, but the result scared him witless. They would need somewhere in the range of 10 million volts.

If you've ever stood anywhere near a 300kV (kilovolt) transmission tower that transports electricity over large distances and heard the occasional pop and crackle, you'd figure that working with this kind of high voltage is quite dangerous. You'd be right, and it was even more terrifying in 1927. We're relatively familiar with electricity today because we use it all the time, but in this period it was still quite new, and voltages this high in a lab setting were unheard of. The risk of a device, operating at millions of volts in the laboratory, sparking and electrocuting Cockcroft and Walton – or more likely Rutherford, who might walk in unannounced – was no laughing matter. What's more, all the pieces of the accelerator would have to withstand the incredible voltage without cracking, exploding or sparking.

While Cockcroft was thinking the problem through, physicists in the United States were already forging ahead with the challenge of generating high voltage. Merle Tuve was trying to use a Tesla coil, like Allibone. Robert Van de Graaff was working on a charge-transferring belt system with a large metal dome. Other attempts were made around same time – pulsed high voltages, capacitor discharges and enormous transformers – all in the name of delivering energy to particle beams. In Europe some German researchers were even risking their lives in the mountains trying to harness bolts of lightning.

Meanwhile, back in Cambridge, Walton and Allibone tinkered away at their attempts to accelerate particles. Walton tried prototypes of a small circular accelerator and a linear version, too, but didn't have success with either. Yet before they could really assess how to proceed, a Russian theoretical physicist named George Gamow arrived in Cambridge and changed everything.

Gamow had recently been in Göttingen, Germany where he had learned about the new ideas of quantum mechanics for his PhD. While everyone else there was studying the electron arrangements in atoms, Gamow decided to try to apply ideas from quantum mechanics to the atomic nucleus. Reading around on the topic, he came upon a recent paper by Rutherford, describing the scattering of alpha particles from a target made of uranium.[4] Rutherford claimed that the alpha particles were scattered as his usual equations predicted, but Gamow wasn't convinced. He happened to know that uranium emits alpha particles in radioactive decay with an energy of about half that which Rutherford was using to bombard it.

Although Gamow didn't know much about the mysterious force holding the nucleus together, he knew it should act the same way regardless of whether the alpha particle was coming into or out of the nucleus. On the way in, as Rutherford was attempting, the alpha particle had to overcome the Coulomb barrier, then would become trapped inside the nucleus. In radioactive decay, an alpha particle had to escape this trapping force first, before the Coulomb force could take over and push it away. The process should be the same in both cases, simply reversed. So how could it be that alpha particles inside the nucleus were somehow cheating and leaking out with only half the required energy?

When he closed the journal, Gamow recalled he 'knew what actually happens in this case. It was a typical phenomenon which would be impossible in classical Newtonian mechanics but was in fact expected in the case of the new wave mechanics'.[5] In wave quantum mechanics, as we saw in Chapter 3, every particle has a wave-like nature that can propagate through space. This means no barrier is 100 per cent solid; the waves can leak through into regions which, classically speaking, they shouldn't be able to penetrate at all. According to Gamow, 'If the wave gets through, even with some difficulty, it always smuggles a particle with it'.[6] We now call this quantum mechanical *tunnelling*. On seeing Rutherford's paper, Gamow quickly formulated a simple model to describe the probability of this happening for uranium and found that his theory explained the radioactive half-life of the element beautifully. Just like that, he had figured out how alpha particles escape the nucleus in radioactive decay. He knew he was onto something.

Gamow then travelled to Niels Bohr's institute, where he did further calculations to figure out whether this idea might also apply in reverse, to help with bombarding elements with artificially accelerated projectiles. Niels Bohr encouraged him to go to Cambridge, but knowing Rutherford's reputation for sometimes ignoring theorists, they planned to smooth his way a little. Gamow arrived in early 1929 armed with a present: two hand-drawn graphs relating to Rutherford's experiments on alpha particle bombardment of light nuclei. The first showed that if the energy of the alphas could be increased, the yield of protons being chipped off light elements would rapidly increase, a tantalising thought to the team stuck counting flashes in the dark. The second showed that for a fixed alpha particle energy, fewer protons were chipped off when the nuclei were lighter. Both of Gamow's theories fitted the experimental data beautifully. The strategy worked, and Gamow was welcomed into the Cavendish.

According to Gamow's recollections he arrived in Cambridge and showed his work to Rutherford, and was then put to work calculating the energy required for protons to get into the nuclei of light elements.[7] With a very simple argument Gamow said it would be about 1/16 the energy of the alpha particles he'd considered previously. 'Is it that simple?' Rutherford asked. 'I thought that you would have to cover sheets of paper with your damned formula.'

Before his visit, one of Gamow's draft papers had already made its way to John Cockcroft, who did a similar calculation. His calculations gave him particle energies in terms of electron-volts or 'eV', that is the amount of energy a particle[8] gains after passing through a potential difference of 1 volt. Thus far he had required protons to reach 1 million electron volts (MeV), which needed a million-volt particle accelerator. He now concluded that there was a small chance a proton with an energy lower than 1MeV could find its way into the nucleus. In fact, the required energy could be as low as 300keV (kiloelectron volts). Cockcroft had already realised the implication of this idea: if protons could quantum mechanically 'tunnel' in through the Coulomb barrier, then perhaps digging into the atomic nucleus could be achieved with a smaller particle accelerator than they thought. The accounts conflict over whether Cockcroft or Gamow was first to make Rutherford aware of the possibility, but the important thing was they both came out with the same result,[9] and they were both in the same laboratory when it happened.

Rutherford's mind was made up. For the first time ever, he would make a major decision based purely on a theoretical prediction because he knew that if they didn't act now, their competitors might beat them to it. He called in Cockcroft and boomed: 'Build me a one-million-electron-volt accelerator; we will crack the lithium nucleus without any trouble!'

Now that Cockcroft needed only a tenth of the voltage he'd previously considered, it started to seem more feasible, and he set his sights on a first attempt at 300,000 volts. This was the minimum voltage, according to his calculation, at which something interesting might happen. But he was desperately busy organising everything in the high magnetic field laboratory next door, so both he and Rutherford realised that he needed a partner adept at running experiments and interested in accelerating particles. They found a willing accomplice in Ernest Walton.

Together, Cockcroft and Walton wanted to build the largest experiment in the whole Cavendish. Even at the new 300kV voltage it was going to be an elaborate, expensive beast. They also recognised that there were challenges they'd need to face beyond the high voltage to make the particle accelerator work. First, they'd need a source of particles. This was easy enough for electrons, but making a steady stream of protons, alpha particles or something else was much harder. They'd need to send those particles through the high voltage to push them up to high energy. Then, they'd have to work out how to control the beam and the operation of the apparatus itself from a safe distance, since it was going to produce radiation. Once they had high-energy particles, they would need to smash them into some kind of target and finally, when all that was done, they'd need a detector system to be able to see what had happened in the reaction.

At least there was one part they weren't worried about. The lab was full of world experts in the use of scintillation counting and new ideas were constantly arising for the detection part of the experiment, including in Wilson's cloud chamber. But they had their work cut out for them when it came to creating the proton source, generating high voltages without destroying the apparatus and managing to successfully control everything.

Putting state-of-the-art equipment designed to carry high voltages into an ill-designed university laboratory was a prospect that would have daunted most physicists, but John Cockcroft was determined to make the accelerator work. Realising they couldn't manufacture everything they needed in-house, he leaned on his former employers at Metrovick, creators of world-leading high-voltage equipment. His first request was a power source for the experiment, a motor generator, which Cockcroft purchased for a good price. Next, they needed a transformer to step up the voltage to 300,000 volts, but when Cockcroft requested one, he ran into trouble. The existing Metrovick transformers used for high-energy X-ray tubes or electrical testing were simply too large to fit through the narrow arched stone doorway of the Cavendish Laboratory. Naturally, Cockcroft asked Metrovick to invent one that would.

The next step was to convert the high-voltage AC from the transformer into a DC source. AC, the alternating current which typically comes from our wall-plug outlets, oscillates between positive and negative values around 50 times a second, but Cockcroft knew this was no good for accelerating particles because the negative part of the AC wave would *de*celerate rather than accelerate the particles. Instead, he needed a DC (direct current) to supply a voltage that would always push the protons down the tube. This meant adding a device called a *rectifier* after the transformer, but no commercial rectifiers were available that could withstand the 300,000 volts they wanted to use. Cockcroft knew that he'd hit a limitation, as it was inevitable that he'd need to go beyond this voltage in the future. So while Metrovick were still working on the new transformer, Cockcroft and Walton set about inventing a rectifier in-house.

Cockcroft was the master of the logistics, but in reality, Walton took on most of the experimental work. One challenge they faced concerned the glass bulbs that made up part of the rectifier. Walton would have glass bulbs made by the in-house glassblower, Felix Niedergesass, then submit them to high voltage using Allibone's Tesla coil, to often disastrous effect. Electric fields concentrate around any sharp edges, whether that's dust or an imperfection in the glass, and the 'corona effect'[10] would cause sparks to creep around the surface and puncture holes straight through it. Getting the shape of the bulbs right took months of trial and error, and eventually the designs became so large that they outgrew Niedergesass's glassblowing lab and had to be ordered from a specialist factory.

In addition to the glass bulbs, they needed specialised wires for the anode and cathode, a heating source for the cathode, corona shields to help prevent sparking and reliable vacuum pumps. Like most of the researchers in the Cavendish, they used red Bank of England sealing wax for all the joints and seals. All the components had to be tested for their ability to withstand the high voltage. Walton tinkered away month after month. He had to work quickly, yet at the same time couldn't rush because he had dangerous high voltages to contend with. Every time a change was needed, he'd have to break all of the wax seals, re-clean, heat and seal it all again before re-testing. As things progressed, he had to spend days hunting down the vacuum leaks and sealing them up.

Rutherford would occasionally drop in on his rounds to see how things were going. He would see the large pieces of equipment appearing from industrial suppliers and complain in his typical style that things were too bulky or too expensive, prompting the Metrovick physicists to say he should be 'made to look at everything down the wrong end of a telescope, since everything is too big for him!' By 1930 Metrovick had delivered on their promise of a new compact transformer designed to fit through the Cavendish door and down the stairs to the basement. Even then, the laboratory floor had to be reinforced to support it. The company also delivered a new vacuum system after one of their scientists, Bill Burch, invented a pump based on a new type of oil he developed (Apeizon). Cockcroft got his hands on a few prototypes before the rest of the world had access.

For all their progress, they had yet to make the proton source or the acceleration tube through which the particles would travel. For the proton source, they tested a range of different setups and eventually settled on a cousin of the cathode ray tube, called a *canal ray tube* or an *anode ray tube*. The device looks similar to a cathode ray tube, a long glass cylinder filled with hydrogen gas with a large voltage applied between the anode (on one end) and the cathode (now in the middle of the tube). Protons are created by the electric field ripping apart the hydrogen gas in the tube and are then pulled towards the negative cathode, which has holes for them to pass through. Finally, they emerge out the other side in the opposite direction of the electrons (cathode rays), creating a lovely fluorescent glow in the tube as they do.

This delicate tube was mounted at the top of the setup so that the protons could travel downwards and into the main accelerating section, a 1.5-metre-long glass vacuum tube. Inside the tube, the high voltage was connected to two cylindrical metal electrodes with a gap between them. The protons would be accelerated by the high voltage as they passed downwards through the gap. The world's first particle accelerator was coming together.

By May 1930 they were ready to run a test. Over the course of a week Cockcroft and Walton slowly brought up the voltage from 50,000V to 100,000V and then eventually to 280,000V, at which point there were signs that they had reached their limit. The beam of protons that emerged, however, was less than satisfactory. It was completely unfocused and spread out across a circle about 4cm in diameter. It wouldn't work to have such a wide beam. To fix that they would have to take the whole thing apart again. But first, they checked to see if anything scientifically interesting was happening. At such a low energy they guessed the protons wouldn't do much to a nucleus, perhaps exciting a few particles inside and emitting gamma rays, so they rigged up a simple electroscope and put a sample of lithium under the beam. Nothing. Beryllium? A tiny effect. Lead? Maybe a little effect there as well, but it might just have been something tricky with the apparatus. Before they could go any further, the transformer failed.

It was time to take stock. With the transformer out of action, they needed to weigh whether it was worth fixing in order to rebuild the 300kV machine. Given the lack of results so far, they weren't sure if it was. What if the calculations they had made were wrong, and 300kV wasn't enough to split the nucleus? Even changing the numbers a little bit gave very different results. Meanwhile Rutherford – now Lord Rutherford – was growing impatient with the lack of results from the test accelerator. They needed to keep him on-side and prove that his investment in their big experiment had been worth it. While it would be quicker to rebuild the 300kV machine than to build a new, bigger version, they had to admit that the 300kV had always been intended as a first step. In the end, room allocations made the decision for them. They were moved into a new larger room, which had light streaming in through beautiful high arched windows along one wall, while the other wall was lined with blackboards. It could accommodate a larger machine easily. Cockcroft and Walton decided they needed to be sure of obtaining results next time, so decided to abandon the 300kV machine and focus all their efforts on constructing a new 800kV machine.

The new design, created by Cockcroft, ingeniously added layers of a voltage doubling circuit to the first rectifier stage.[11] With four of these steps, they could take a 200kV input voltage up to 800kV. They replaced the hard-to-make bulb-shaped tubes for the rectifier and accelerating section with more reliable glass cylinders instead, after coming across the idea in the work of Charles Lauritsen, a physicist at Caltech in the US. They also discovered that rather than using wax to seal joints, plasticine did a much better job and was much quicker to re-seal if things needed adjusting. As before, Walton worked tirelessly to build the new machine, somehow writing up his PhD thesis in the meantime.

In early 1932, almost four years after Cockcroft and Walton started, a major new discovery was made in the Cavendish. It didn't, however, belong to them, but to James Chadwick. He'd been quietly experimenting in the background for years when he caught wind of results from Irène Curie and Frédéric Joliot in Paris showing alpha particles from a polonium source striking beryllium and supposedly producing a gamma ray with an inconceivably high energy. He knew their experiments were right: they were exceptionally thorough in that regard. But he disagreed with their interpretation. In just a few weeks, he performed a new series of experiments demonstrating that the bombarded beryllium emitted not gamma rays, but a neutral particle roughly the same mass as the proton. After a quest lasting almost twelve years, Chadwick had finally discovered the neutron.

With this new breakthrough Rutherford was losing patience for the hugely expensive and time-consuming accelerator project. Legend has it that when he went to check up on them, Rutherford walked into their lab, hung his wet coat on a high-voltage terminal and promptly electrocuted himself. Recovering from the shock, he lit his pipe in a burst of ashes and smoke and told them to get on with it.

On the morning of 14 April 1932, Walton was alone when he had finished getting the improved machine to warm up. Cockcroft had run off to look after something in another lab. At Rutherford's insistence they had set up his favourite detector, a zinc sulphide screen, instead of an electroscope. Walton placed a target of lithium at the bottom of the accelerator tube and stabilised the machine with a voltage of around 250,000V. Next, he adjusted the settings to bring the proton beam on. Curious to see if anything was actually happening, he crawled across the floor from the control station to the accelerator, avoiding the high-voltage components, and clambered into the lead-lined box they had built for observations. He pulled the black cloth to shut out daylight, adjusted the microscope and peered through it.

Bright flashes were appearing all over the screen. Walton hadn't spent long in the nursery lab, but he could immediately guess what he was seeing: alpha particles. Too many to even count. He ducked out again and turned the beam off. They were gone. He switched it back on, and there they were. He could barely believe it was real. He fetched Cockcroft, who quickly repeated the tests. Together they called Rutherford, bundled him into the tiny box with his knees up around his ears and showed him what they'd found. Sure enough, they were alpha particles – and since he'd been the one to discover alpha particles, he should know! Chadwick later agreed. They knew without even having to spell out to each other what had happened: the protons were getting into the nucleus of lithium, atomic number 7, and the nucleus was splitting off into two alpha particles. They'd artificially caused a nuclear disintegration for the first time in history.[12] What's more, they'd managed it with protons at roughly 250keV, far lower than the 1MeV or even 10MeV they'd expected. Gamow's quantum theory was right.

They swore each other to secrecy until Cockcroft and Walton had made the necessary checks and written a quick paper to send to *Nature*. While they did this, for a week in the spring of 1932, only four people in the world knew that the atom had been split. They continued with experiments at a frantic pace, placing stacks of thin foils in the path of the alpha particles to confirm that they were emerging from the nucleus at a tremendous speed. Each alpha was zipping along with 8MeV of energy, which at first sounds impossible given that the protons only came in with a few hundred keV, but this measurement bolstered their confidence that their understanding was correct. The mass of the combined proton and lithium before the reaction was ever so slightly higher than the mass of the two alpha particles afterwards. Taking this mass difference and converting it to energy using Einstein's already famous equation $E=mc^2$, their calculations accounted for the 8MeV energy almost exactly.

Rutherford invited Cockcroft and Walton along to a meeting of the Royal Society on Thursday 28 April. The crowd had gathered to celebrate Chadwick's neutron discovery and Rutherford mentioned this great achievement in his prepared opening speech. Then he stayed at the podium. After a dramatic pause he announced that two young men present in the auditorium, John Cockcroft and Ernest Walton, had artificially accelerated particles and successfully split the nucleus of lithium and a range of other light elements. He had only to raise his hand to gesture at the two young men for the audience to burst into spontaneous cheering.

Within days, newspapers announced 'Science's greatest discovery'.[13] The news spread rapidly throughout the world, the *New York Times* leading with 'The Atom is Giving Up Its Mighty Secrets'. Cockcroft and Walton quickly adapted to a new life of posing for the cameras with Rutherford or with their equipment, looking slightly embarrassed as they were interviewed by the sudden flood of journalists at the laboratory door.

Their competitors were left pinching themselves. If they had known that they could split an atom of lithium with as little as 125,000V,[14] they could have been the ones to make the discovery. Even Cockcroft and Walton might have succeeded two years earlier if only they'd put a zinc sulphide screen in the experiment – so that they could easily see each alpha particle flash – instead of an electroscope. When they finally did so, it turned out to be much easier to work with than the more abstract movements of the leaves in an electroscope. They simply hadn't quite believed that the low voltage of their first accelerator would be sufficient. By the end of 1932, other labs around the world were working hurriedly to convert any devices with sufficient voltage to atom-smashing mode. A whole new field of nuclear physics was born.

Rutherford and his team had finally not just one, but two new ground-breaking discoveries almost at the same time. The existence of the neutron was finally confirmed, but far more exciting was the ability to artifically split the nucleus of an atom in two. Rutherford had achieved his goal of understanding what was inside the nucleus: protons and neutrons. The experiment also confirmed the importance of quantum mechanics in the nucleus and that Einstein's $E=mc^2$ held true when splitting atoms. They were firmly back in the lead of the race to understand the nucleus of the atom, and now, for the first time, Rutherford and his team had the ability to break nuclei apart at will in order to study them further. Rather than relying on cosmic rays, they could now control their experiments by changing the type of particles they were accelerating, how many were used and their energy, to study the impact they had on any sample they wished to bombard. They could switch them on and off whenever they liked. The nucleus was theirs to explore.

With the sudden ability to artificially accelerate particles, demand from researchers for accelerators rapidly increased. Companies were quick to act in taking on the new technology, often for use in their own research labs. In Europe, Philips in the Netherlands built rectifiers and whole Cockcroft-Walton generators, as their design came to be known, and even later sold one to the Cavendish when they expanded their high-voltage lab in the mid-1930s. Their competitors in the United States, including Van de Graaff, also found commercial success with their high-voltage accelerators. Soon after the big discovery, Westinghouse started building high-voltage machines using Van de Graaff's method and by 1937 had built a 5MeV accelerator known as the Westinghouse Atom Smasher. By the mid-1950s, any self-respecting university physics department or lab had to have a particle accelerator. Today, a handful of companies still make this type of machine and you'll find their products in research institutes and labs worldwide.

If you've ever seen one of these devices, you'll never forget it. In the north of England is the Cockcroft Institute, which today specialises in designing and building new particle accelerators.[15] In the institute's huge brightly lit atrium, visitors are stopped in their tracks by the sight of an enormous metal device. Four dark brown ribbed ceramic insulators reach up, circled around halfway up by doughnut-shaped metal rings, as reddish copper tubes zig-zag between them. The whole structure climbs up to the three-storey ceiling and culminates in a huge, bulbous silver metal terminal. This particular Cockcroft-Walton generator formerly provided protons for a large accelerator facility[16] at the Rutherford Appleton Laboratory, just south of Oxford. While this generator provides a great first impression, it is not actually all that old: it lived a life of reliable service from 1984 until 2005, when it was finally decommissioned, and replaced by more modern technology.[17]

It was not a Cockcroft-Walton but a Van de Graaff-type accelerator that nuclear physicist Harry Gove had in his lab at the University of Rochester when Charles Bennett first asked him to help find the date of the violin in 1977. This seemed impossible, at least until the idea arose to use the accelerator to detect very small traces of carbon-14. For their first experiment they bought bags of hardwood barbeque charcoal from local shops to represent present-day carbon (from trees that had recently been cut down). They inserted it into the ion source, the starting point of the accelerator which vaporises samples and uses a high voltage to strip electrons away to create a beam of charged ions that can be accelerated downstream. As a comparison, they also found a sample of graphite, sourced from oil deposits which are millions of years old and in which the carbon-14 should have reduced to nothing. On 18 May 1977 they ran the two samples and found that the charcoal had over 1,000 times more carbon-14 compared to the graphite. As Gove recalls, 'It was one of those instantly recognisable triumphs that occur all too infrequently in science'.[18]

Instead of using the traditional method of radiocarbon dating, which required a large sample or a very long time, the particle accelerator enabled them to take a tiny sample and accelerate all the individual atoms and isotopes. Once at high speed, the particles would be bent using a magnet, and since the carbon-14 would naturally curve slightly less than carbon-12 because of its heavier mass, the relative amounts could simply be counted using a detector. The particle accelerator provided exquisite control and precision, allowing them to circumvent the natural limits on radiocarbon dating. It quickly became clear that the potential applications were enormous.

Meyer Rubin, a geochemist who led the carbon dating unit for the US Geological Survey, saw the article and quickly got in touch with Gove and his team. Rubin said he had been quietly hanging onto piles of small geological samples that were too small for the traditional carbon dating method, waiting for someone to create a way to measure them. A few weeks later he arrived in Rochester to work with Gove and Bennett's team to try measuring milligram-sized samples using the new accelerator technique.

Rubin was excited about the potential for measuring small samples, particularly in geology, climatology, oceanography and dendrochronology (the study of tree rings). Together, the team produced a series of breakthroughs using the new technique: they checked the method by dating organic samples back 48,000 years, finding they agreed with Rubin's earlier measurements, which had used much larger samples. Collaborating with the many researchers who contacted them, the Rochester group dated Antarctic meteorites, ice, a woolly mammoth and even ancient air samples containing not milligrams, but just micrograms of carbon-14, all with success. In 1978 Rubin brought in a piece of cloth wrapping from an Egyptian mummy that had been estimated at around 2050 years old, and together they ran an experiment which verified the result. Then they turned to a compelling yet controversial request.

Around 1979 the team were contacted by the British Turin Shroud Society, with the idea of putting a date to the artefact that Jesus was supposedly buried in. It took ten years to come to fruition, leading to a famous investigation in 1987. Small samples were sent to a range of labs around the world that had adapted or installed particle accelerators specifically for this purpose, including Rochester and the radiocarbon dating facility in Oxford. Gove and Rubin found it was medieval (AD 1260–1390) with a 95 per cent confidence level and not 2,000 years old at all. The other labs all confirmed Gove's results. Despite the results, the Turin Shroud is still revered today.

The technique that emerged, which Gove is (in part) credited with inventing,[19] is called Accelerator Mass Spectrometry, or AMS. Today there are facilities in labs which use dedicated particle accelerators for this technique not just in the United States, but in Turkey, Romania, Australia, Japan, Russia and China to name just a few. Many of the countries which host these facilities are interested in understanding their own rich geographical and cultural histories, and AMS provides the possibility of piecing together the stories of rare and precious objects without destroying them. Just like with Bennett's violin, the samples required for AMS are at least 1,000 times smaller than for traditional carbon dating. In most cases, there is no other definitive method to establish chronology. Accelerator technology has since opened up new insights into history, geology, archaeology and many other fields.

Bennett, it seems, never found out if his violin was a real Stradivarius or not, or at least never authenticated the unlikely claim, as nothing more has been recorded about it.[20] But by then, perhaps he had forgotten it altogether, caught up in the ultimate scientific excitement of inventing the most accurate method of dating historical artefacts that we know of.

Today, most people still assume that particle accelerators and the beams they create are only ever used by physicists and come nowhere near our food, water, household goods or our own bodies. Yet from the chips in our phones and computers, to the tyres on our cars, to the shrink wrap on our food, we are surrounded every day with objects that have been strengthened or otherwise improved using particle beams. Often these particle-based methods of irradiation or modification are chosen because they're quicker, more environmentally friendly and more effective than using alternatives like chemicals or manual processing. This is not a small market: it's estimated that around $500 billion-worth of products are formed or modified using particle beams each year in the United States alone. Many of these machines are electrostatic accelerators, descendants of the one Cockcroft and Walton used to split the atom in the early 1930s.

One of the largest applications is in the semiconductor industry. The powerful computer chips inside our smartphones and laptops are based on electronic components made from semiconductors, which form the 1s and 0s that all computer logic is based on. To make a semiconductor like silicon into a useful device it needs to be made slightly impure by adding dopants: tiny amounts of other elements like boron, phosphorus or gallium. It's these dopants which give precise control over the electrical properties of the semiconductor, but they are very difficult to add using chemistry. The only precise way to do this is to control individual ions and implant them using a particle accelerator, a process called ion implantation. Without particle accelerators in these factories we wouldn't have modern semiconductor-based electronics which are now embedded in digital cameras, washing machines, televisions, cars, trains and even rice cookers.

It's not just semiconductors that can be modified with particle beams – even jewellers are in on the act. The diamond company DeBeers has accelerators producing ion beams which they use to bombard raw gemstones. This can change the colour of a diamond, or transform turquoise from a dusky pink colour into the transparent blue that it is celebrated for.

Meanwhile, just fifteen metres underneath the famous glass pyramid, the Louvre in Paris has a particle accelerator dedicated entirely to art. The facility is called AGLAE – the Accélérateur Grand Louvre d'analyse élémentaire – a 37m-long installation which is used to bombard artefacts from the museum to find out which elements they are made of. Under the guidance of the director of the laboratory, Dr Clare Pacheco, the team use the accelerator for a range of applications that are collectively called ion beam analysis. One of the techniques they use regularly is known as Rutherford Backscattering Spectrometry (RBS). They count ions bouncing backwards from a target, looking for exactly the kind of result that the Cavendish scientists used with gold foil to show that the atom had a nucleus. Now, with the controlled conditions of an accelerator, they can utilise the full power of this idea in a way the original experimenters would never have dreamed. The sample artwork is placed in the line of the particle beam, and a detector picks up the ions that scatter backwards. For each detector position, different atomic nuclei will bounce back different numbers of ions, and the accelerator changes the ion beam energy to map out a characteristic curve of energy versus number of ions. Then it's just a matter of comparing the curves to those of known materials to figure out which atoms are in the sample, and their relative amounts. This technique has been used, for example, to confirm that a scabbard belonging to Napoleon was really solid gold. With this and other methods, Dr Pacheco's team can identify even the slightest trace of elements across the periodic table, from lithium to uranium, helping to uncover the secrets and origins of artwork and historical artefacts without inflicting any damage. If you've ever wondered how art historians know unequivocally that an artwork is authentic, this is one way.

These same techniques are used to measure the exact composition of the glass of bottles of antique wine and compare them to known authentic bottles. Wine fraud is a large and growing problem in the fine wine industry. In one case, a collector spent \$500,000 on four bottles of wine that were said to have belonged to former US president Thomas Jefferson. Ion beam analysis of the bottles found that they were fake, and a lawsuit was launched against the wine dealer pretty quickly.

The same idea is also starting to be used in forensics. Most methods to pick up traces of drugs like cocaine or of residues from gunshots destroy the samples. But scientists including Dr Melanie Bailey at the University of Surrey, UK, are now using ion beam analysis to study evidence found at crime scenes.[21] Without destroying the evidence, she can check the elemental composition of a sample and find tiny amounts of drugs or residue that other methods might miss. She can even compare her results with material found on the clothing or body of suspects: tiny soil samples picked up on a shoe can place a suspect at the scene of the crime.

For the physicists of 1932, all of these applications were far off in the future. Cockcroft and Walton worked with the accelerator for a couple of years, but soon new researchers took the lead. John Cockcroft took over running other parts of the laboratory and later worked on the use of nuclear power for peacetime energy supply. Ernest Walton took an academic position back in his native Ireland at Trinity College Dublin. This intense period in their careers, which won them the Nobel Prize in 1951, would never be repeated.

Their success – which came in the same year that the positron was discovered – realised Rutherford's dream of discovering what was inside the nucleus. The various pieces of the puzzle now came together: nuclei contain both protons and neutrons, usually in roughly equal numbers. Isotopes vary in mass because they have differing numbers of neutrons, while the number of protons stays the same. Some configurations are more stable than others, and the unstable ones are radioactive. Rutherford's quest was now to understand the *forces* that somehow held the nucleus together. How did the presence of neutrons manage to prevent the positive protons from pushing the nucleus apart? This motivated the idea of a new, *nuclear force*, holding them together.

While Cockcroft and Walton's invention was still being exploited both scientifically and industrially, it became clear that particle accelerators using enormous voltages would soon reach a fundamental limit. A new technology was needed. Little did they know that that very technology, already developed in the United States, had almost beaten them to their world-famous result.

6

Cyclotron: Artificial Production of Radioactivity

In 1932, the year a particle accelerator first successfully split an atom, the list of fundamental particles found in nature was growing quickly. It included the electron and its antimatter version the positron, along with protons and neutrons. These were all thought to be indivisible, although we'll see later that protons and neutrons also have structure. Photons, particles of light, had now been introduced, and just four years later positive and negative muons, the heavy cousins of electrons and positrons, were found. No one knew the significance of the particles that weren't part of atoms and whether they were important or whether there were more out there like them. All they knew was that to find out more, they would have to follow the lead of Cockcroft and Walton and smash atoms.

There were hints that nudged them in this direction, one of which we've already seen – the fact that some unknown force seemed to bind protons and neutrons together inside the atom and prevent them from flying apart. The other hint came from chemistry, or to be precise, what was missing from chemistry. Uranium was the heaviest known substance in the periodic table[1] at this time, but there were four gaps corresponding to elements 43, 61, 85 and 89. Arranging the elements by atomic weight and with similar chemical properties stacked in columns, Russian chemist Dmitri Mendeleev had predicted in the nineteenth century that these elements should exist, along with others that had subsequently been found. For example, underneath aluminium in the table was a gap where Mendeleev predicted an element which he called 'eka-aluminium' that should have a particular density, melting point and chemistry. Gallium (element 31) was subsequently discovered in 1875, matching his predictions almost exactly. With the benefit of hindsight it is easier to call the missing elements the names we now have for them, technetium (43), promethium (61), astatine (85) and francium (89), but in the early 1930s they had never been observed and so were nameless.

You might think scientists would be out looking for these missing elements, but they weren't really expending much energy in that direction – and for good reason. The discovery of radioactive elements had taught them that not all the elements in the periodic table were stable, as chemists had always assumed, so it was possible the missing elements had simply disappeared over time and thus wouldn't be found. Now that radioactivity was involved, the atom was turning out to be an unpredictable and confusing place, one which was dynamic in ways that chemistry could not reveal. The larger goal lay in understanding the nature of atoms and the structure of the nucleus, along with the forces that held all of matter together. This meant exploring and understanding the details of as many elements as possible, attempting to build up a comprehensive theory that could predict the properties of elements and their isotopes, known and unknown, radioactive or not.

If only they could create beams of particles sufficiently powerful to break apart atoms of every element, who knew what they might accomplish? This is what had driven Cockcroft and Walton to tame enormous voltages and build the world's first particle accelerator, but they were not the only ones working on the problem. Within a few years their work would be overtaken by a young American named Ernest Orlando Lawrence. The machine he invented would not only end up dominating the field of nuclear physics, but would also bring scientists from different disciplines together to work across boundaries and open up unexplored areas. As a result, Lawrence's work would also transform medicine for ever.

Lawrence never intended to be a physicist. He was determined to study medicine when he enrolled at the University of South Dakota and chose chemistry as a supporting major. His love of physics first had to be coaxed into being by a mentor.

It was Ernest Lawrence's hobby that drew him into his mentor's orbit. Growing up in South Dakota, Lawrence and his neighbour Merle Tuve spent most of their spare time building radio equipment, communicating via Morse code in the Tuve family attic, learning about and installing relays, transmitters and other bits and pieces. When Lawrence went off to university he left his radio equipment at home, but soon found himself wishing the university had its own set. Lawrence tracked down the Dean of Electrical Engineering, Lewis Akeley, and presented to him a clear, coherent argument for purchasing some radio equipment, including lists of parts with their prices.

That evening, Akeley went home to his wife and enthused about Ernest Lawrence, about his scientific curiosity and clear ability. But why wasn't Lawrence enrolled in either physics or electrical engineering? Why was he studying medicine and chemistry? Convinced that Lawrence had genius in the area of physics, he granted him $100 to buy the radio equipment, gave Lawrence a space to put it and left him in charge. Akeley, a physicist by trianing, was careful not to push Lawrence to change course, as he believed good students would realise the value of physics by themselves. He tentatively asked Lawrence if he thought physics would be helpful given his interests in wireless, but Lawrence didn't think so. He'd studied a little bit of it in high school, but doubted he had the ability to accomplish anything in the subject.

Despite himself, Akeley invited Lawrence for dinner and started regaling him with tales of great physicists and their adventures: from Heinrich Hertz, who saw the connection between light and electricity and became the first person to transmit waves by wireless, to Marie Curie and her discovery of radioactive elements. Most thrilling were the tales of Ernest Rutherford, who had shown that the atom wasn't solid after all. Akeley laid out the adventure that awaited explorers in the field. They explored the inner world of matter and unlocked the secrets of the Universe on its smallest scales, upon which everything else, including Lawrence's beloved chemistry, biology and medicine, was built. A well-trained mind, Akeley insisted, gave you the ability to learn in any field, and physics could provide that training. He gave Lawrence a final proposition: if he'd spend a month over the summer learning physics with him, and was still not interested, Akeley would never speak of the idea again. Lawrence agreed. By the time the other students returned, the bet had paid off.

'Class. This is Ernest Lawrence,' Lewis Akeley announced in his physics lecture one day. 'Take a good look at him, for there will come a day when you will all be proud to have been in the same class as Ernest Lawrence.' They stared at the tall young man with his charming smile, neat light brown hair and blue eyes. One day, when Lawrence fell asleep in class, Akeley simply told the rest of the students 'Never mind. Let him be! He knows more physics asleep than the rest of you do wide awake.'[2] Akeley can't possibly have known what lay ahead, but his words would turn out to be prophetic.

By 1928, at the age of just twenty-seven, Ernest left to become an associate professor at the University of California. Finally in charge of his own research programme, with the freedom and encouragement of a young institution behind him, all he needed was a good research question to investigate.

By this point in our story, we have a head start on Lawrence because we already know where things stood in 1928, and what was to come in just a few years. We know that Gamow's theory would spur Cockcroft and Walton to develop their accelerator in Cambridge. We know that only a few hundred keV of energy is sufficient to split a nucleus of lithium. But Lawrence, like Cockcroft and Walton at the time, knew none of this. He knew physicists had found electrons and X-rays and that the atom had a nucleus, and he was aware of the counter-intuitive realities of quantum mechanics and wave particle duality. He knew that cosmic rays were continuously bombarding us from space, and that C. T. R. Wilson's invention of the cloud chamber allowed us to study them, though Lawrence wasn't particularly interested in detectors at the time.

While many scientists were studying cosmic rays, it seemed to Lawrence more important than ever to be able to control high-energy particles in the laboratory. He wasn't satisfied with the attempts being made. His old school friend Merle Tuve was trying to tame voltages up to 1MV, while Cockcroft and Walton and their competitors were doing the same, but Lawrence wanted to know where research would go after 1MV was achieved. He had a whole career ahead of him; he didn't want to set out on a path that would fizzle out after just a few years. To Lawrence there seemed to be a fundamental flaw in the idea of using high voltages to accelerate particles. Even if they could create a usable voltage of a million volts, they still wouldn't be able to get particles beyond the 5MeV energies of alpha particles emitted from natural sources (like radium), since the high voltage translates directly into particle energy. A million volts could give a million eV (1MeV), but could never give 5MeV. If the mysteries of the atom were ever to be revealed in the laboratory, someone would have to come up with a practical method to reach high energies, tens or hundreds of MeV, without the corresponding high voltages.

In 1929, Lawrence was reading journals late at night in the University of California library. On a whim he picked up a journal on electrical engineering written in German and flicked through until some diagrams and equations in an article by a Norwegian named Rolf Widerøe captured his attention. Lawrence didn't speak German, but the idea was clear enough to understand.

Lawrence himself later described the idea as being so simple that even children could understand it intuitively: when you sit on a swing there are two ways to get it to swing high in the air. You can either give the swing one almighty push, or you can give a series of small pushes at the right time, building up the oscillation using the concept of resonance. The existing ideas for accelerators took the first approach, but Lawrence realised that the second method was what he needed to do. Instead of using a single very high voltage to accelerate particles, Widerøe's diagrams seemed to suggest applying an oscillating voltage to a series of metal tubes lined up end-to-end, with spaces or 'gaps' between them. The voltage on the tubes would flip from positive to negative and back again a few million times a second, at a reasonably modest voltage. Particles would pass through the middle of the metal tubes as if through a water pipe, and only in the gaps between tubes would the particle 'see' the voltage.[3] With the right timing, the particles would get a little kick forward at each 'gap' just like each small push on a swing. A series of tubes all powered by the same oscillating source would only need a modest voltage, but the overall energy gained through a series of these tubes could be very high.

Widerøe's idea was good, except for one fundamental flaw. To reach high energies the line of tubes would have to be incredibly long. What if, instead of having many tubes in a row, Lawrence could bend the particles around in a circle and re-use the same accelerating 'gap' many times over? He could use the resonant acceleration concept to create, as he called it, a proton merry-go-round.

In his haste to check if his idea could work, Lawrence grabbed a paper napkin and started scribbling the equations down. He knew he'd be able to bend the particles using a magnetic field, using the long-understood fact that the force from a magnet can push on particles at right angles to their direction of travel. Each revolution, the particles would gain a little bit of energy, spiralling outwards into larger circles as they went faster. In working through the equations, he realised that the greater speed of the particles on each larger circle would compensate exactly for the longer path they had to follow, so the time taken to return to the voltage gap would remain the same on every turn. That meant he could use a voltage that oscillated with a constant frequency, which would be easy to engineer. It was almost too good to be true.

He rushed back to the faculty club and asked the first mathematician he could find, Donald Shane, to do a quick check of his calculations. Shane confirmed his maths was sound, then looked at Lawrence and said: 'But what are you going to do with it?'[4] 'I'm going to bombard and break up atoms!' replied Lawrence.

It was such a simple and elegant idea that Lawrence wondered why it had never been thought of before. Despite his excitement, he didn't immediately start building anything as he had already made plans to travel across the country. He went to Washington for the Physical Society meeting, then to Boston to see his brother John before going on to General Electric in Schenectady, New York, where he'd promised to spend two months. Along the way he gave talks and dined with many top physicists, including Robert Millikan. Everywhere he went he would tell whoever would listen about his new idea.

Most of them could think of a reason why his idea would not work. They said there was no way to focus the particles in such a device, so it would never be able to attack something as small as an atomic nucleus.

They thought the particles would never stick to the spiral-shaped path or would fly off vertically and smash into the chamber and be lost. They wondered how Lawrence planned to get particles out of the machine, although on this point at least he had some ideas. Even his old friend Merle Tuve expressed doubts, while Lawrence for his part looked sceptically at Tuve's use of a big Tesla coil to accelerate particles. But by the time Lawrence returned to California, he was ready to test his idea.

Lawrence's first PhD student at the University of California, Niels Edlefsen, was six years Lawrence's senior and just finishing his thesis. It was 1930 and Edlefsen hadn't yet decided which job he would take after his PhD, so found himself with a little time to spare. Edlefsen wanted to focus on theoretical work and brush up for the examination at the end of the doctorate, but Lawrence had other ideas. He insisted to Edlefsen that his radical new idea for a particle accelerator was much more exciting than studying theory, and that he couldn't see any reason why it wouldn't work. Edlefsen couldn't see anything wrong with it either and after studying for two more weeks, he eventually relented and agreed to give it a go. 'Good!' Lawrence said. 'Let's get to work. You come up with what we need right away.'[5]

In spring of 1930, Edlefsen started with a glass flask about the size of a perfume bottle that he flattened and coated with silver. He carefully scraped back a narrow strip of silver across the middle, leaving two separate silver regions for electrodes. The flask could be evacuated of air and had apertures for the ion-producing filament, the introduction of proton-producing hydrogen, and an electrical probe to determine results. All the apertures were then sealed with wax. Lawrence, in the meantime, had done some smooth talking to wrangle permission to use the biggest magnet in the department. The idea was that the flask would be wired up, pumped down to vacuum and placed between the poles of the magnet, which would make the particles spiral around in circles as they gained energy. Finally, they were ready to put the big idea to the test.

They turned it on. The glass cracked. The glass chamber clearly wasn't going to work. Undeterred, they came up with a new idea. They took a small round copper box, which Edlefsen cut in half to form the electrodes. These were then fixed to plate glass with wax, the two halves of the box separated by a small distance with the openings parallel to one another to form two 'dees' (so called because they take the shape of two capital letter 'D's). If you imagine taking a big cookie wrapped in copper, then breaking it down the middle and removing the cookie, the two halves of copper left are what the 'dees' looked like. A radio-frequency oscillator was connected to the dees to produce an alternating voltage. It looked a bit of a mess. After all of Lawrence's talk, the other members of the lab didn't hold back in teasing Edlefsen and Lawrence about their supposedly powerful machine for accelerating particles.

Whether or not Edlefsen successfully managed to accelerate protons in the device is unclear. He did get some protons at least circulating around, but before he could produce any definitive results, he had to move on to the job he'd lined up elsewhere. But to Lawrence, the project was promising enough that he immediately put a new student to work on the resonant accelerator.

That student, Milton Stanley Livingston, was the serious-looking son of a minister who had traded chemistry for physics at university. The only son of the family, he had grown up on their farm in California among tools and machinery, which taught him practical skills about designing and building complex systems. Those skills were now going to be put to the test, as he worked on the device which would come to be known as the 'cyclotron'.

Livingston put together a tiny device that could fit in the palm of your hand and was similar to Edlefsen's attempt, though a little neater. It was just 11cm in diameter, made of brass and sealed with wax, and cost about $25 to make. Livingston was quick to make progress and over the Christmas break of 1930 he and Lawrence used this 11cm model and an oscillating voltage of 1,800V to accelerate protons up to 80,000eV, showing that the concept worked. The cyclotron could accelerate particles up to energies that were many times greater than the applied voltage, just as Lawrence had dreamed up in the library.

They tweaked the designs as they built, learning everything along the way through trial and error. They altered the shape of the electrodes and the size of the gap between them, and adjusted the magnet slightly to achieve focusing, massively increasing the beam current. A few weeks later they built a cyclotron just shy of 30cm in diameter, for which they had an even larger magnet built. When they tuned it up, Livingston found they had produced protons racing around at just under a million eV, by applying just 3,000 Volts. Lawrence literally danced around the lab: finally, his machine could smash atoms.

Once again Lawrence was due to travel, and while he was away expounding the virtues of his new invention which had almost – but not quite – reached the magical million-volt mark, Livingston kept at it. On 3 August 1931 Lawrence received a telegram to say that the record had finally been achieved: 'Dr Livingston has asked me to advise you that he has obtained 1,100,000 volt protons. He also suggested that I add "Whoopee!"'.

Lawrence had been visiting his girlfriend, Molly Blumer, when the news came through. He read out the telegram to her family. While they were still congratulating him, he escorted Molly outside and proposed to her. She accepted, on the proviso that she finish her own studies at Harvard before the wedding. Then he rushed back to the lab and spent the subsequent days with Livingston demonstrating the invention to any colleagues and friends who wished to see it. With a relatively tiny and inexpensive machine, they had surpassed the energy achieved by Cockcroft and Walton's room-sized generator.

If at this point they had actually done what Lawrence had aimed to do all along – smash atoms – then the history of nuclear physics would look somewhat different. Instead, his team of about ten physicists and engineers were determined to reach higher and higher energies. Spurred on by Lawrence's infectious enthusiasm, they built larger machines, first a 27-inch (69cm) cyclotron aided by the donation of a large magnet from the Federal Telegraph Company, which was quickly redesigned as a 37-inch (94cm) version. Before long, they had reached 2,000,000eV protons.

Why didn't they use the cyclotrons to do science? Why did they get so caught up in building bigger and bigger instruments? In succeeding to build the cyclotron they had actually invented a whole new field of physics, my own field, accelerator physics. They realised that the control and manipulation of beams of charged particles was a fascinating area of research unto itself and that making progress in this area would secure future progress in physics, as Lawrence predicted would happen if researchers were limited by high voltages. In successfully accelerating beams with the cyclotron they had already confounded their many detractors who had said it wouldn't be possible. Now they had to work on understanding exactly how it worked, and how to improve it, which required detailed knowledge of the physics and behaviour of charged particles. They were pushing so far beyond the limits of technology that they were creating new knowledge in our understanding of physics and engineering: of how beams of subatomic particles create and interact with electric and magnetic fields, how to design electromagnets with precise properties, and how to focus, transport and measure beams of subatomic particles that are invisible to the eye.

Lawrence and Livingston's enthusiasm for perfecting the machine led the team to miss a number of important discoveries. In 1932, just as the cyclotron was winning the high-energy race, they were – scientifically speaking – left in the dust by those with simpler experiments. Chadwick discovered the neutron and measured its mass to be very similar to the proton. At Columbia University, Harold Urey discovered a new isotope of hydrogen, with a single charge but double the regular mass, called deuterium. In the same year Anderson used a cloud chamber to discover the positron. Then the big news came through in April: Cockcroft and Walton had successfully smashed atoms for the first time. Lawrence's team quickly set up the cyclotron with a lithium target to reproduce the same results. In just a couple of weeks they had easily extended the proton on lithium results up to 1.5MeV, almost twice the energy the Cavendish could reach. As per Gamow's quantum tunnelling theory they found that the higher energies increased the rate of reaction even further. Even if they weren't first past the post, at least they were right in believing that high energies would allow them to more effectively smash atoms. Now, with the highest energies in the game, they were off and racing.

The cyclotroneers, as they became known, decided to create an experiment no one else could. They got the university chemistry department to produce some deuterium, or 'heavy hydrogen'. They put it in their ion source to strip off the electron and produce deuterons (deuterium nuclei) to use as projectiles in the cyclotron. With one proton and one neutron, they figured the heavier deuterons would be a more potent way to penetrate the nucleus than protons. By 1933 they were in a territory of their own and producing results that were quite bewildering. All the elements they bombarded with deuterons seemed to produce enormous reaction rates, far higher than what they could accomplish with protons. The reactions would always produce neutrons and protons with surprising amounts of energy. The only conclusion, according to Lawrence, was that the deuteron was breaking up. If that were true, he calculated that the neutron might be far lighter than Chadwick had measured.

Before he could figure it out, an invitation arrived. Lawrence was invited to attend the 1933 Solvay Conference in Brussels, the meeting of the great and the good in nuclear physics. At first Lawrence didn't plan to attend because of his heavy teaching load, but the invitation was such a great honour for his lab and the university that they allowed him to skip some teaching and agreed to send him over first class on a ship. In preparation, Lawrence pulled together all the results he could on the deuteron experiments.

In Brussels Lawrence found himself among every big name in physics, from Albert Einstein to Marie and Irène Curie and of course Lord Rutherford. When it was his turn, Lawrence talked about the great promise of the cyclotron and presented his results from the deuteron experiments. Far from making the great impression he'd intended, however, most of the others were sceptical, or at best they thought he must have had made a mistake. Rutherford, now the self-described grandfather of nuclear physics, agreed with them. Despite that, he took a liking to the outspoken trailblazer. He nudged Chadwick, who mustn't have been too impressed with the young American, and said 'He's just as I was at his age!'

Afterwards, the Cavendish team showed, using their Cockcroft and Walton accelerator, that the deuterons were forming a layer of heavy hydrogen on the surface of the target. The reactions Lawrence's team had seen were deuterons hitting other deuterons, not the target elements disintegrating. This explained why the results looked similar for every target material, and on calculating the neutron mass with the correct reaction in mind, the mass of the neutron turned out to be safe. Lawrence, chastised, wrote to all concerned to apologise for the mistake. To his team, he insisted that 'science can grow through mistakes too', but he had now learned his lesson. They would have to be much more careful in the future.

*

Part of the reason Lawrence and Livingston kept missing out was their lack of particle detecting and counting devices – the one thing that definitely wasn't lacking in the Cavendish Laboratory. Lawrence's team had tried to develop a Geiger counter, but abandoned two attempts as the counters seemed to be plagued by high background counts. They didn't have cloud chambers either, so their measurements were quite basic even though the cyclotron could produce far higher energies than other machines.

After the Solvay Conference and the deuteron fiasco were behind them, Lawrence and Livingston got back to work, as did all their competitors in labs around the world. In 1934 Lawrence came running into the lab waving a copy of a French journal. When he had caught his breath, he told his team the news: Irène Curie and Frédéric Joliot in Paris had induced radioactivity by using natural alpha particles on targets of light elements. They didn't even use an accelerator.

Realising they had all the elements of an artificially produced version of the same experiment in front of them, Livingston recounts they: '... turned the target wheel to carbon, adjusted the counter circuits, and then bombarded the target for 5 minutes [...] The counter was turned on, and click-click--click---click----click. We were observing induced radioactivity within less than a half-hour after hearing of the Curie-Joliot results.'[6]

Lawrence's team were so focused on developing the cyclotron technology that they had also missed out on being the first to detect artificially induced radioactivity. On this occasion at least they were not alone, as the Cavendish and every other lab with an accelerator had also missed it. They had wired up their Geiger counter to the same switch as the accelerator, so as soon as the beam was switched off, the counter was, too. If they had left it on, they would have realised from their very first experiments that the cyclotron had been making radioactive elements. At least they could now understand the reason for their inability to create a reliable Geiger counter: the whole laboratory was radioactive.[7]

With the results from Curie and Joliot, Lawrence realised that dozens of new radioactive elements could be made. Using the cyclotron, they could bombard different elements with protons or deuterons, change the neutron and proton number and produce radioactive isotopes. Now they could go beyond the naturally occurring radioactive elements. They could re-create the reactions in stars that made these elements in the first place. Perhaps, they could even create elements and radioactive isotopes that were no longer found on Earth, or had decayed away to very small amounts.

A less driven team, with a less inspiring leader, would have been disheartened that their cyclotron was beaten to first place in the race to split the atom by Cockcroft and Walton – by just a few weeks – and then in recognising the production of artificially induced radioactivity. Irène Curie and Frédéric Joliot were awarded the Nobel Prize in chemistry for their discovery just one year later. But if he was envious of others' success, Lawrence didn't show it. 'There are discoveries enough for everyone', he would tell his students.[8] Besides, he wouldn't trade places with Cockcroft and Walton or with Curie and Joliot, because now he had created a machine that could overtake them all.

Within a day or two of the Curie-Joliot result in 1934, Lawrence had discovered radio-sodium[9] by bombarding deuterons onto a target of sodium chloride (table salt). The cyclotron could produce millions of radio-sodium atoms per second, which would then decay away with a half-life of 15.5 hours emitting electrons and gamma rays. Once again he found that the higher the energy of the cyclotron beam, the higher the yield of radio-sodium. Radio-phosphorus followed quickly after. We can only imagine the excitement he must have felt, knowing that as he built higher energy machines the world of radioelements would open up before him. Dozens, if not hundreds, of new radioactive substances might be found. In the midst of that excitement, it occurred to him that, perhaps, these new radioactive elements might turn out to be useful to society.

Lawrence wrote to his younger brother John, who was by this time a medical doctor specialising in haematology. In summer 1935, John Lawrence arrived at the 'Rad Lab' – as the Radiation Laboratory was known – on holiday from Yale, encouraged by Ernest to see what he could do with the new radioisotopes in the medical field. X-rays were already known to have the potential to kill human cells, offering a potential future treatment for cancer, but no one had yet tried using radioisotopes. Because the new isotopes had the same chemical properties as their non-radioactive counterparts, John realised the systems of the body might treat radioactive elements the same way it did normal ones. Salt made of radioactive sodium would be processed the same way as regular salt, for example. He could then use the radioactive properties to interact with the body, or perhaps even to detect the radiation emerging to see the processes inside organs, without having to make a single incision in the skin.

John made a start using radioactive phosphorus-32 produced by the cyclotron to investigate the metabolism of animals. Phosphorus is the second most abundant mineral in the body after calcium, making up 1 per cent of bodyweight, and is involved in the formation of bones and teeth among many other functions. He prepared a group of mice with leukaemia and injected them with radioactive phosphorus, then left to go fishing on the local river. Two weeks later he returned to find that the group of mice he had injected were alive and in apparently good health, whereas all the 'control' group mice, who hadn't been injected, were dead. Within months he was trialling the radioactive phosphorus on human patients, with impressive results – the phosphorus helped with remission of their disease.

A little later, John and Ernest together tested what would happen to a rat if it received radiation externally. They placed it in the cyclotron in the beam chamber which sat between the upper and lower magnet poles, next to a beryllium target, and switched the beam on at a very low dose. After about a minute, John called for the cyclotron to be switched off so he could check on how the rat was doing. It was dead. This terrified the cyclotron team, who became fearful that the biological effects of radiation might be far worse than they knew. They worked to update the shielding around the cyclotron. Later, John realised the rat had died not of radiation, but of suffocation: it had been placed within the vacuum vessel and all the air had been taken out for the experiment. Despite that, there was suddenly a lot of interest in the effects of radiation on people, both positive and negative.[10] The experiments were promising enough that the next year John moved to the University of California, set up his own lab and team, and the two brothers worked together for many years.

If you had walked through the Rad Lab in those days, it would have been a very crowded place. In the same space, there were cages full of mice, wet labs for chemical separation and electrical apparatus for the physicists, not to mention the cyclotron and its shielding. You'd have been among not just physicists, but experts from many fields including engineers, chemists and biomedical scientists among others. Lawrence couldn't always afford to pay them – many joined purely out of enthusiasm for the work. The new medical applications helped him a lot with attracting money, which was particularly important during the Depression. The radio-sodium work had been achieved with a 27-inch cyclotron producing 6MeV deuterons with a relatively modest current, but in 1937 the cyclotron was upgraded to a 37-inch machine, double the current and an 8MeV beam energy. With this, the medical researchers had enough radio-sodium and radio-phosphorus for their work, while the physicists also had an energetic enough beam to do detailed work in nuclear physics.

On a typical day, the cyclotron would bombard a target, which would be handed to someone from the chemistry department, who would do the chemical separation work. This usually involved dissolving the target and then distilling it to separate out chemicals based on their boiling point. Sometimes the separation of the dissolved elements would require other techniques, like adding additional chemicals to force the element to reform into a solid, or separating out elements using chromatography. Once that was done the physicist would take over again using an electroscope or other instrument to measure the activity and half-lives of the products. Using this method, the chemist Glenn Seaborg in 1937 discovered a new radioactive isotope of iron, Iron-59, which became immediately useful in the study of blood diseases.

John and Ernest saw the greatest potential in the direct applications of radiation to cancer treatment. They did experiments using neutrons that showed some early promise. They also studied high-energy X-rays produced using a linear accelerator built by Lawrence's colleague David Sloane. In 1937, John and Ernest got the news that their mother had uterine cancer and she was given just a few months to live. The clinic she was at – Mayo clinic – didn't want to treat her with radiation, but the brothers took matters into their own hands, asking one of the doctors working with John to help treat their mother using X-rays. As John Lawrence later said in an oral history interview, 'to make a long story short, this massive tumour just started evaporating.' She was about 67 at the time, and lived until she was 83. We will come back to the concept of radiotherapy in much more detail in Chapter 10.

In 1938 Seaborg discovered Cobalt 60, an intense gamma ray emitter with a half-life of 5.3 years, which later found immense application as a radiation source, at its peak usage delivering 4 million therapeutic irradiations annually in the United States alone. It is still used widely in medicine and in industry as a well-regulated radiation source.[11] In the same year, a discussion with a doctor made Seaborg aware of studies in thyroid metabolism using iodine-128, which had a half-life of twenty-five minutes – so short it was limiting the studies. The doctor said he would prefer one with a half-life of around a week. Seaborg and his colleagues quickly identified iodine-131 which had, coincidentally, a half-life of around eight days. There was such rich ground for discovery with the cyclotron that it was almost as if they could invent new isotopes on demand. Iodine-131 is now used millions of times a year for diagnosis and treatment of thyroid disease, diagnosis of kidney and liver disorders, and for functional tests of organs. Seaborg's own mother was treated with iodine-131 and her life was extended by years as a result.

As the number of medical uses grew, physicists kept on pushing the limits to find new radioelements and to piece together what they were learning about the structure of the nucleus and the many ways it could fit together. They could create not just radioactive isotopes of known elements, but also elements that had never been seen in nature before, to fill in the missing gaps in the periodic table. The first completely new element, found in 1937, was technetium (atomic number 43). It was isolated by Emilio Segrè in Italy after he visited the Rad Lab and convinced Lawrence to mail him a thin foil of molybdenum that had been part of the cyclotron to see if he could determine what type of radioactive elements were present. After a series of chemical separations and purifications, Segrè and his colleague Carlo Perrier found evidence for two isotopes of technetium: technetium-95m (with a half-life of six days) and technetium-97m (half-life of ninety-one days).

Every isotope of technetium is radioactive, and since the predominant naturally occurring isotope technetium-99 has a half-life of

211,000 years, it is very difficult to find in nature because practically all of it has decayed away during the lifetime of the Earth.[12] But with the cyclotron it was easy to create. In 1938 Segrè moved to the United States and collaborated with Glenn Seaborg using the cyclotron to confirm the existence of another isotope of the new element, technetium-99m. This isotope has a shorter half-life, about six hours, and is a step in the decay of a technetium nucleus during which gamma rays are emitted.

Technetium-99m turned out to be an incredibly important isotope for medical diagnosis, and was first used to take a medical scan of the liver in 1963. By the end of the 1990s, it was being used in more than 10 million diagnostic procedures per year in the United States alone, imaging the function of the thyroid, brain, liver, spleen and bone marrow among other parts of the body. Demand has only increased, and it remains the most widely used medical radiotracer worldwide. Seaborg and Segrè had no premonition of its potential application in medicine when they did the work.[13]

The other three missing elements of Mendeleev's periodic table were filled in over the next few years. All four of the missing elements turned out to be radioactive, which explained why they went undetected because there was very little of them left naturally on Earth. The longest-lived isotope of francium-233 has a half-life of just twenty-two minutes (discovered in 1939 by Marguerite Perey in Paris), while astatine-210's is 8.1 hours (discovered in 1940 by Corson, Mackenzie and Segrè in California) and promethium-145's is 17.7 years (discovered in 1945 by Marinsky, Glendenin and Coryell in Tennessee). Once the periodic table was full, the cyclotron allowed the Berkeley physicists to push beyond it. Over the years, Seaborg and other physicists would create heavier and heavier elements, driven by the question of how many neutrons and protons can be held together in a nucleus and in which circumstances they are stable or unstable. Seaborg was awarded the Nobel Prize in chemistry in 1951 for his discovery of the transuranic elements plutonium, americium, curium, berkelium and californium. Seaborg and his Berkeley colleagues would later go on to synthesise einsteinium, fermium, mendelevium, nobelium and of course, seaborgium, named after him.

Thanks to the cyclotron and other accelerators, the periodic table has massively expanded from the time when uranium (atomic number 92) was the heaviest known element. Today the heaviest element made in a laboratory is ununoctium (118), also known as oganesson after its discoverer, Yuri Oganessian. It was produced in 2016 in Dubna, Russia and so far, only four atoms of it have been made, so its chemical and physical properties are still being understood. Research into the formation of superheavy elements is crucial to understanding the creation of heavy elements in the early Universe, and still continues at many labs around the world.

The periodic table shows elements arranged by atomic number, or number of protons, but after the enormous expansion of radioisotopes with the cyclotron there now exists a second version, the 'chart of nuclides' – also known as the Segrè chart – where the number of neutrons is plotted on the horizontal axis and the number of protons on the vertical axis. The stable elements of the periodic table lie on a roughly diagonal line, but around them is now drawn a wide band of exotic and unstable nuclear configurations known as *nuclides*, arranged and coloured by the type of radiation they emit when they decay.

At Berkeley, as the cyclotrons became more and more powerful, a new laboratory was funded and opened in 1939. The Crocker Laboratory housed a 60-inch machine, and Lawrence's team now numbered sixty people to build and run the cyclotrons. On occasion they consumed so much power they would cause blackouts in the local town. Somehow, among all this frenetic work, Lawrence found time to visit Stockholm to receive the 1939 Nobel Prize in physics. Discoveries kept on coming, too, including the discovery of carbon-14, the isotope which is key to carbon dating. As tensions rose around the world in 1939 and 1940, Lawrence planned and built an even larger machine designed to surpass the 100MeV energy barrier for the first time. Achieving this higher energy required a much larger magnet to confine the beam. To double the energy, the weight of the magnet had to increase by a factor of eight, and it would need as much iron as a warship. The enormous 184-inch machine, the pinnacle of cyclotroneering, was constructed at a new site up the hill from the original Rad Lab, and it was still under construction when World the Second World War broke out.

Many physicists, including Lawrence, were recruited during the war to understand how energy could be released from nuclei to be utilised in a weapon and the huge new cyclotron was commandeered for the war effort. Meanwhile, John Lawrence contributed by developing imaging techniques that used radioactive gases to study the internal functioning of the human body. Working together with Cornelius Tobias, one of Ernest Lawrence's students, he used radioactive isotopes of the gases of nitrogen, argon, krypton and xenon (produced using the 60-inch cyclotron) to discover the nature of 'the bends', or decompression sickness. This was in the days before aviators wore pressurised suits. Today, radioactive krypton gas is still used in hospitals to take images of patients breathing.

The most likely place you'll find a cyclotron today is not in a large laboratory, but in the basement of a hospital. More than fifty types of radioisotopes are now created and commonly used in medicine and almost all major hospitals have a nuclear medicine department. There are radioisotopes that can treat diseases and diagnose when our hormones, blood flow or other organ functions aren't working properly. If you ever need an 'image' taken of the function of your thyroid, bones, heart or liver, chances are you'll be using a technique which originated with the Lawrence brothers and their team. Around the world, about 15 to 20 million of these 'scans' are performed every year – about one per hundred people in developed countries.

Without the collaboration between John and Ernest Lawrence, without the quest to smash atoms with bigger and bigger particle accelerators and without interdisciplinary collaboration, none of this would have happened. Seaborg later said that when he was working to identify radioisotopes, he had no inkling of the ultimate beneficial clinical applications. Lawrence certainly had no intention of building a machine that would transform medicine. John and Ernest, when they were young, had no intention of working together in the way they did. Yet Lawrence and his laboratory would later be seen as pioneers in multidisciplinary collaboration, and the founders of the era of Big Science.

The inspiration that Lawrence had to make a circular accelerator paved the way for higher energies than could ever have been achieved before. For decades the cyclotron was the machine of choice for nuclear physicists. Even Chadwick built one at the University of Liverpool, enlisting Lawrence's help and telling him it was one of the most beautiful instruments ever invented. Yet for all the discoveries and advances in medicine, the energy of the cyclotron beams was still far less than the energy of the particles coming from cosmic rays, and eventually, even these beautiful machines started to reach their limit.

The enormous amounts of iron required for the magnets made it increasingly difficult to build larger machines. Even if the iron was available, they knew that the laws of physics would eventually scupper plans for larger and larger cyclotrons. Einstein's special relativity dictated that as particles approached the speed of light, they would continue to gain energy but would no longer gain speed. This meant that with increasing energy, the particles in the cyclotron would fall out of synch with the accelerating kicks and languish at an upper limit of perhaps a few hundred MeV. Something would have to change.

7

Synchrotron Radiation: An Unexpected Light Emerges

In 1933, Bell Labs radio engineer Karl Jansky was scanning the sky in 'short wave' or radio frequencies using an antenna. He was trying to figure out if there were any sources of noise that would interfere with AT&T transmitting telephone signals across the Atlantic. Instead he found a mysterious hiss of what he poetically called 'star noise', cosmic radio waves which were strongest in the direction of the edge of our own galaxy. For millennia, humans had been looking up at the night sky without knowing they were only seeing a fraction of what is out there, because our eyes can't see beyond the visible spectrum. Jansky's discovery showed that much of the light coming from the Universe isn't in the visible spectrum at all, but in the radio spectrum.

It was a coincidence that this discovery happened at the same time that nuclear physicists were learning about nature on the smallest scales. The two areas – astronomy and nuclear physics – at first seemed unrelated, until a serendipitous discovery using particle accelerators led to the bridging of knowledge between the two. The result was not just a greater understanding of astrophysics, but powerful instruments that are now used across almost every domain of science, the findings of which have impacted all our lives.

At first Jansky's cosmic radio wave discovery was ignored by the astronomy community. But another radio engineer, Grote Reber, soon followed up on it. Reber funded and built the first radio telescope in Illinois in 1937, and found bright sources of radio waves in the constellations Cygnus and Cassiopeia. In time astronomers began to pay attention and this new tool led to a remarkable shift in our view of the cosmos. By the 1950s and 1960s radio astronomy had opened up a view of the Universe we'd previously been blind to. Celestial objects all around were emitting radio waves, including our galaxy the Milky Way. Jesse Greenstein, an astronomer, later described the dawn of radio astronomy for the *New York Times* as having 'led to information that overturned the idea of a rationally developing universe [...] and replaced it with a relativistic, ultra-high-energy cosmos of scary, violent, uncontrollable forces like black holes and quasars. It was a revolution.'[1]

Radio astronomy led to many discoveries, like that in 1945 by geologist and physicist Frances Elizabeth Alexander which established that radio signals were coming from the Sun. In 1967, Jocelyn Bell-Burnell found objects emitting intense regular pulses of radio waves resembling an extraterrestrial beacon, earning them the nickname 'little green men'. Properly named, 'pulsars' are extremely compact rotating stars that emit radiation from their poles, from which astronomers have learned much about the processes at the end of a star's life. The discovery of pulsars was considered so important it was awarded the Nobel Prize, but not to Bell-Burnell – apparently due to her status as a PhD student at the time. Instead, the prize went to her supervisor Antony Hewish.[2]

Today, much of what we now know about cosmology, black holes, supernovae and other spectacular objects in the Universe comes from decades of work in radio astronomy, but back in the 1940s there was still a big question to be answered: *how* do these objects out in space, from pulsars to our own Milky Way, emit radio waves? The answer would come from back down on Earth, from the physicists building accelerators to dig into the atom.

At the beginning of the 1940s a new type of particle accelerator arrived on the scene, which became known as the 'betatron'.[3] The suffix 'tron' means an instrument, while 'beta' radiation is made up of high-energy electrons, exactly what scientists wanted to produce artificially with the new machine.

Why not just use a cyclotron? It turns out that they are great for protons and deuterons, but cyclotrons aren't very good at accelerating electrons. The cyclotron, as we saw in the last chapter, is a machine that relies on a magnetic field to bend charged particles around in a circle and an oscillating electric field that pushes the particles up to speed. As the lightest members of the particle world, electrons very easily reach speeds near that of light, and relativity dictates that while they can gain more energy at those speeds, they no longer get faster. This means the oscillating electric field acting on the cyclotron 'dees' loses synchronicity with the electrons and starts decelerating them instead. Physicists in search of high-energy electrons for generating X-rays or scattering experiments were stuck. The idea of the betatron showed, as Lawrence was fond of saying, that there's 'more than one way to skin a cat'.

Betatrons are designed to work in a rather different way from cyclotrons. They use the principle of magnetic induction: that is, they use the idea that a changing magnetic field induces a current in a circle of wire, the same way that an induction cooker generates a current that heats up a frying pan. A beam of electrons travelling in a circle can act as if it were in a wire or frying pan – so putting electrons in a changing magnetic field can give the beam energy, and contain and focus it at the same time, with no concerns about the timing of oscillating voltages. This idea was actually the same one the young Ernest Walton had pitched to Rutherford in the late 1920s. Walton's attempts to make one work back then were unsuccessful, which was part of the reason he ended up building an accelerator with John Cockcroft.[4] Although his early experiments failed, Walton made key contributions to the theory of such a machine, including figuring out how to make the particles stay on the desired orbit. This is actually more difficult to achieve than you might think.

In a circular accelerator the aim is to keep particles orbiting perfectly on a ring-like trajectory that goes around indefinitely inside a circular tube – known as a 'doughnut'.[5] When working with a real beam of particles, we have to think of them not one by one but as a collection of independent particles, each of which is never perfectly in the middle of the tube. Instead, each particle follows its own trajectory that is not quite on the ideal orbit. Walton was rightly concerned that as particles were accelerated, they would need to be constantly pushed back to the centre of the tube so that they didn't fly off and get lost. He made a detailed calculation of how to do this, shaping the magnetic field to slowly decrease as the radius increases so the field would bulge around the outer edge of the ring. This setup, he found, focuses the particles and ensures they are always pushed back to the ideal orbit.[6]

By 1940 the first working betatron was finally realised by Donald Kerst in the United States, and the new machine quickly became a promising technology for accelerating electrons up to around 99.99 per cent of the speed of light. Now that electrons could be accelerated, a use was quickly found not just in science, but in the real world too. In particular, a growing market emerged for particle accelerators in medicine and industry. In 1944 physicist Herb Pollock led a team at General Electric Research Laboratory in Schenectady, New York, to build a betatron designed to reach energy levels of 100MeV. The ribbed iron front of the 130-ton machine rose far above the heads of the physicists, looking more like a battleship than a medical device, with the 'General Electric' label riveted across the magnet. A gap in the iron at around head height made space for the ring-shaped vacuum vessel. The machine made a deafening whirring noise when it operated as large electrical currents swirled around and through the electromagnet coils, accelerating beams from zero to 100MeV sixty times a second.

The director of the GE research lab was physicist and engineer William Coolidge, whose plan was to use the betatron to create high-energy X-rays by smashing the 100MeV electrons onto a target, creating a super X-ray tube whose rays could go right through the body or industrial objects for imaging, where lower energy X-rays might be stopped. He hoped it would become a commercial device, after which the team would build larger and larger betatrons as the market grew. The best part was that they saw no limit to the electron energy they could reach with such a device.

Just as they got used to operating the machine, John Blewett, a physicist in a different group at GE, became aware of a theory that seemed to pose a problem. Working in the Soviet Union, Dmitri Ivanenko and Isaak Pomeranchuk had pointed out in a letter to *Physical Review* that there was a problem with accelerating electrons in a circular machine. If you apply the principle of conservation of momentum to a charged particle being bent around a corner, you find that this bending must cause the particles to emit radiation.[7] Blewett repeated the calculations and realised that the Russians were right.

For the 100MeV betatron, the effect was predicted to be small. The energy lost would be just 10eV per revolution, so that the final energy of their machine would be 99MeV rather than 100MeV. They could cope with that. But the calculations predicted that for every doubling in electron energy, the energy loss would increase by a factor of sixteen. If they wanted to create larger betatrons, vast amounts of radiation would be emitted as the particles got to higher energies. So much energy would be lost, said Ivanenko and Pomeranchuk, that the acceleration mechanism wouldn't be able to keep up. The best upper limit, they said, was a particle energy of about 500MeV. If it were true, the betatron idea would soon be obsolete.

Some of the GE team were sceptical that such an effect would exist. After all, electrons travel through wires all the time and don't emit radiation. Blewett insisted on running a test at GE to see if the predictions were true. They had the 100MeV betatron at their disposal and it was important to rule it out. Blewett calculated that the orbit should shift a little due to the radiation effect.

When they turned the machine on and made measurements, the orbit did seem to be a little off. But then again, it was a complicated machine and an orbit shift could happen for a number of reasons. The 'smoking gun' would be the radiation itself. They placed equipment around the machine to search the radio spectrum for emissions, but found nothing.

The question was still unsolved in late 1945, when Ernest Lawrence paid one of his regular visits to Schenectady and shifted their attention towards a new goal. In a seminar he presented an idea that his team at Berkeley was working on. Instead of the spiralling particles in the cyclotron, he proposed a machine with a beam constrained to a single orbit, where acceleration would be provided by radio frequency electric fields and the strength of the magnets would change in time with the acceleration. The idea had recently been invented simultaneously by Lawrence's Berkeley colleague Ed McMillan and by Vladimir Veksler in Russia, fleshing out an idea that Australian Mark Oliphant,[8] one of Rutherford's students, had come up with a few years earlier. This new concept would do away with the need for the giant magnets of cyclotrons and betatrons, but the trade-off would be a slightly more complicated operating principle: as the speed of the particles changed orbit by orbit, the accelerating frequency would have to vary in time to keep up. Everything would have to be perfectly synchronised to make it work, which gave the machine its name, the *synchrotron*.

At this point the GE physicists would have been listening intently. They already had a betatron, but were worried about the technology reaching an upper limit in energy because of radiation losses. The synchrotron idea seemed interesting, but how would it fix the problem? How would the synchrotron continue accelerating electrons to higher energies once they were radiating?

McMillan and Veksler solved this problem with the principle of *phase stability*, which relied on the timing of the radiofrequency fields used to accelerate the beam orbit by orbit. It's easiest to imagine a bunch of charged particles in a circular accelerator as a group of surfers riding a (voltage) wave. If a slower surfer needs to speed up, they can move up the wave where its curve is steeper; if they need to slow down, they can move to the bottom of the wave. By arranging the timing correctly with respect to the voltage wave provided by the radiofrequency fields, the front (faster) particles see a lower voltage than the rear (slower) particles and stay bunched.

This not only keeps particle bunches all grouped and accelerating nicely, but McMillan also claimed it would overcome any energy loss from radiation. This would just be like surfing into a headwind: the surfers would all need to move up the wave a little to keep going, but they could do so provided the wave was high enough.[9] The upshot was that the synchrotron could exceed the 500MeV energy limit predicted by Ivanenko and Pomeranchuk.

The attraction for Lawrence was clear, as the synchrotron idea seemed scalable up to an almost unlimited energy, unlike the cyclotron he had invented. He was determined to build a synchrotron to get to high energies and also to avoid the increasingly enormous amounts of iron his cyclotrons were using. In typical Lawrence style, however, he had not yet built one: he was just telling everyone about it while he and McMillan worked on plans. For the GE physicists his seminar made two things suddenly clear. First, the reign of the betatron might be even shorter lived than they had imagined – the synchrotron might be the path to super high-energy electrons instead. Second, they might be able to build a small synchrotron before Lawrence built his to demonstrate the idea for the first time in the world.

The GE physicists immediately got approval to build a 70MeV synchrotron and began designing the different parts. The magnet itself weighed 8 tons and had a 2.5-inch gap in the middle for a 70cm-diameter circular glass 'doughnut' through which the beam travelled.[10] They designed a clever power circuit that would transfer energy around to raise and lower the magnetic field at the right time in order to control the particles. Meanwhile Blewett, who had moved on from GE, had left them some calculations he'd received from the respected theorist Julian Schwinger which gave a few further insights into the radiation predicted by Ivanenko and Pomeranchuk.

Schwinger would later share the Nobel Prize with Richard Feynman and Shinichiro Tomonaga for developing quantum electrodynamics (QED) in the late 1940s. Schwinger's calculation said that the radiation given off from a circular trajectory wouldn't be emitted in all directions; instead, it should form a tight beam pointing forward along the path of the particle. He predicted that the frequency of the radiation would shift higher as the electron energy increased. Finally, he noted that at the energies the GE team were working with, the radiation should extend beyond the radio frequency range, up into visible frequencies.

The synchrotron started operating in October 1946,[11] but it wasn't running as smoothly as they'd hoped. Components kept failing and having to be replaced, but they kept at it and in April 1947 things were running pretty well, except for one problem: they kept seeing sparking in the machine. Technician Floyd Haber was sent in to observe the synchrotron while it was operating to try to see what the problem was.

It would have been dangerous to stand near the machine and watch it while it was running, so Haber rigged up a large mirror, measuring 6 x 3 feet, to watch the machine while safely hiding around the corner of the thick concrete wall for protection. As the scientists pushed the limits of the machine, Haber called out that he saw sparking and told them to shut it off. Normally, if there's sparking, the vacuum level – the pressure in the 'doughnut' – changes rapidly, but this case was different: the vacuum level remained stable. One of the physicists, Robert Langmuir, came to watch as well, and together they observed a small, very bright bluish spot coming from the synchrotron.

Langmuir immediately realised what he was seeing. He called for someone to stop accelerating the beam and the light went away. It just had to be 'Schwinger radiation'. Amazed that their electron beam was producing visible light, the team quickly decided to test the prediction that the colour of the light should be related to the energy of the particles. As they adjusted the energy downward they observed – with what must have been a mixture of satisfaction and disbelief – the spot of light change colour from blue, to yellow and then to red before it disappeared entirely. The whole thing, as one of the team later recalled, took about thirty minutes.[12] By sheer luck the new vacuum chamber was made of glass, so they could see the light coming out of the machine from the circulating electrons. They had simply missed out on seeing the effect three years earlier in their betatron because the metal beam chamber had blocked the light. It was one of those rare moments of serendipitous discovery, one which would go on to have a major impact.

Light emitted in this way is called *synchrotron radiation* and has very specific properties. It can be incredibly intense, it's coherent (laser-like rather than light-bulb-like) and it can cover the full electromagnetic spectrum, from X-rays through visible light to infrared, depending on the magnetic field and the electron energy. The light is also *polarised*, that is, the oscillations of the light waves all take place in the same direction. Light can become polarised in many ways, including when it reflects from water or a car bonnet, which polarise it mostly in the horizontal direction. This is why polarised lenses in sunglasses block out glare by only letting the vertically oscillating light waves through.[13] Synchrotron light emerges polarised in a direction related to the electron bending: for a beam circulating in an accelerator, it comes out polarised in the horizontal direction. Its properties are unique enough that you can use measurements to determine when it is being produced: if you measure light with the right properties you can deduce that it is almost certainly coming from electrons bending in magnetic fields.

This insight turned out to be the key to solving the astronomers' questions of what was producing radio emissions out in space. The

Milky Way, pulsars and many other objects aren't just balls of gas and dust: they also have magnetic fields. When charged particles are bent in those fields, they emit synchrotron radiation just like in the accelerator, lighting up the Universe, usually in the radio-wave spectrum. Astronomers can check if the emissions are polarised and from that figure out the magnetic structure – the location and strength of magnetic fields – of objects out in space.

As radio astronomy grew in the 1950s and 1960s it emerged that magnetic fields are far more common than had previously been imagined. One spectacular example is the Crab Nebula, the remnants of a cataclysmic supernova in AD 1054 in the constellation of Taurus, which turns out to have an energetic cloud of electrons spiralling around magnetic field lines driven by a pulsar at its centre. We now know that stars, galaxies, neutron stars and supernovae all have magnetic fields. Magnetism may also explain the behaviour of the most extreme objects in space, including the enormous jets of ionised matter ejected from supermassive black holes, thought to be caused by particles accelerated in the tangled magnetic fields at the centre of these dense compact objects. With an understanding of synchrotron radiation, astronomers were able to use the radio emissions they were detecting from space to gain insights into objects like this, revealing the magnetic properties of our Universe.

At GE, the light was at first a curiosity, which the physicists showed off to all their visitors. Then they figured out they could use the light to help adjust, optimise and operate the synchrotron, which helped them design their next machines to sell. Over the next few years higher energy synchrotrons were built around the world and soon it became apparent that synchrotron light had potential beyond just diagnosing the electron beam. The betatron inventor Donald Kerst put it best when he remarked, 'Wouldn't it be interesting if these beautiful and sophisticated machines made their greatest contributions to science as light bulbs?'[14] In many ways, Kerst's ironic prediction was accurate. Once it could be produced in the lab, synchrotron light would go on to become an unbeatable tool in scientific research, from chemistry and biology to materials science and archaeology.

The first scientists to attempt to use synchrotron radiation were at Cornell in 1956 and, five years later, at the US National Bureau of

Standards – which sets agreed units to help with work in fields like radio, the automotive industry and electronics. This confirmed that synchrotron light is far superior to any standard light source or X-ray tube. Others quickly followed by adapting existing synchrotrons to allow users to access the light for experiments. At first these side-users had to vie for time and space at nuclear physics facilities, but by 1970 the first user facility was built: the Synchrotron Radiation Source (SRS) at Daresbury Laboratory in the UK. Governments around the world started building particle accelerators not for atom-smashing physics, but to meet the demands of a vast range of scientific and commercial users. By 1974 there were more than ten synchrotron facilities worldwide designed and built specifically to create synchrotron light.

Images can be made using synchrotron light by placing samples in the light downstream of a window or port in the vacuum chamber and recording the result, originally using photographic plates, as was done in the 1970s, and nowadays with digital detectors. The samples being studied can be incredibly diverse: examples include chocolate, steel and even pieces of sea cucumber.

One field that has benefited perhaps more than any other from synchrotron light is the field of structural biology. Biology, it turns out, depends ultimately on physical structures at the microscopic scale: the way that proteins fold, diseases take hold and even the very structure of DNA. As Oxford biologist Professor David Stuart explained in an interview for the Nuffield Department of Medicine, structural biologists work to understand biology in a very detailed way, rather like figuring out how a car works by studying each individual piece, how it fits with other pieces and how they work as a machine. Biology is more complex than a car, though. Organisms like us are composed of trillions of cells, which have a remarkable variety of inner components that operate at the nanoscale. When we understand how biology works at this scale, it gives us the ability to intervene in a very precise way when something goes wrong.

Our current understanding of structural biology relies on the real jewel in the crown of imaging techniques, called X-ray crystallography. It has been in use since well before synchrotron light sources existed, and no less than twenty-eight Nobel Prizes have been awarded based on it. It started with father and son British-Australian physicists William and Lawrence Bragg at the University of Adelaide in 1913, who took an X-ray source and fired it at a salt crystal. It produced a beautiful diffraction pattern which they realised could tell them about the structure of the crystal itself, all the way down to the level of atoms.[15] In their wake, scientists have refined this technique to unravel the structure of just about every important molecule and material. Kathleen Lonsdale (William Bragg's colleague) used X-ray crystallography to figure out in 1929 that benzene is a flat ring, while Dorothy Hodgkin unravelled penicillin (1949), vitamin B12 (1955) – an achievement that won her the Nobel Prize in 1964 – and insulin (1969), the latter task taking her thirty-four years. In 1952 Rosalind Franklin famously used X-ray crystallography to produce Photograph 51, showing that the structure of DNA was a double helix. The structures of graphite, graphene, haemoglobin and myoglobin, among countless others, have been determined this way, all done just using conventional X-ray tubes. But with the advent of synchrotron light sources, crystallography became vastly more powerful and remains so today.

Synchrotrons enabled enormous breakthroughs in fundamental science. Using crystallography Sir John Walker and others worked out the structure of adenosine triphosphate (ATP), the molecule which transports and stores energy in all plant and animal life, including humans. Roger Kornberg figured out how genes copy themselves using mRNA, and Venki Ramakrishnan and colleagues worked out the structure of the ribosome, all Nobel-winning discoveries. Note that none of these breakthroughs were in nuclear or particle physics, the fields which led to the serendipitous discovery of synchrotron radiation in the first place.

At first this added scientific power might seem far removed from everyday life, until you realise that our understanding of the basic biology of viruses also depends on X-ray crystallography. This suddenly became of urgent importance when when COVID-19 first emerged in Wuhan, China at the end of 2019. Inside the SARS-COV-2 virus are twenty-eight proteins. These proteins are chains of molecules which are folded up on themselves in very precise ways – imagine a ball of wool in an intentional tangle. This folding process leaves so-called 'active sites' which can be targeted with chemical compounds. Structural biologists can replicate these proteins for study by using their genetic structure to clone them. But first, someone has to sequence the code of the virus.

After the virus first emerged in China on 29 December, it took just twelve days before six virus sequences were made available. By 5 February 2020 Zihe Rao and Haitao Yang's team at ShanghaiTech deposited the first structure of the main protease (a protease chops up proteins, but is also essential for viral replication and an attractive target in drug discovery) into the Protein Data Bank, the online repository that scientists around the world use as a central location for their data. They had solved the structure using the Shanghai Synchrotron Radiation Facility. By then the team had already proactively shared the information with over 300 research groups worldwide.

Before most governments had taken any action, structural biologists were hard at work at synchrotron light sources around the world creating and studying the physical structures of the proteins that make up SARS-COV-2. That's because they knew that for a drug or vaccine to be effective against a virus, the human body has to produce molecules that physically recognise, attach to, and then neutralise and eliminate the unwanted pathogen. Any treatment or vaccine option has the same starting point: they need to know how the virus works. The key to that knowledge lies in the virus structure and function. Once we understand the chemical basis for the body's recognition of a virus we can try to design a drug to reduce its effect or a vaccine that can tell the human body to create antibodies. Perhaps counter-intuitively, one of the key front-line battles in the COVID pandemic was fought not in hospitals, but at ring-shaped buildings the size of a football field containing machines that come from the realm of particle physics.

At the Australian Synchrotron, half an hour outside Melbourne, Dr Eleanor Campbell works as a beamline scientist, an expert who conducts experiments with synchrotron light and helps other scientists do the same. While everyone else she knew was sent home to work when the pandemic hit, Campbell was one of the few scientists whose work went into overdrive on the site. She looks after a beamline called 'MX-2' used for 'macromolecular crystallography', which allows scientists to figure out the arrangement and shapes of biological molecules down to the atomic level. In normal times, her beamline users work in chemistry, condensed matter physics, engineering and Earth and materials sciences. During early 2020 it was entirely devoted to COVID-related studies.

The beamline receives synchrotron light from the heart of the facility, the synchrotron itself, hidden behind big walls of concrete shielding. The main ring is made of a repeated pattern of electromagnets – shoulder-high blocks of iron powered by thick copper cables – which is fed high-energy (3GeV) electrons by a smaller accelerator. A specialised operations team runs shifts to keep things going 24/7. The electrons inside the synchrotron can be kept circulating and emitting light for days or weeks, pumping out radiation while the energy is continuously replenished. When one bunch of electrons is retired and dumped out of the machine, another quickly takes its place so there is very little gap in radiation for the users.[16]

A series of experimental stations appear at tangents round the circumference of the ring. These are the beamlines. Their location is dictated by the 'insertion devices' placed around the ring to generate synchrotron light. Nowadays, rather than just using the radiation emitted naturally in bending magnets, 'insertion devices' called *wigglers* and *undulators* literally wiggle the beam around, producing beams that can be tuned to a specific wavelength. The light then travels through a window or port which opens up onto the beamline, where scientific users set up their experiments, placing their protein crystals in the sample holder, ready to collect data.

The first step is successfully making a protein into a crystal, one of the most difficult parts of the work. Biological molecules are large and pliable – in other words, squishy – while what we think of as crystals, like salt, are traditionally rigid. Campbell's work involves convincing 'a mélange of biological matter to start forming an ordered, rigid crystal'. This is a process of trial and error testing many reagents – starting with chemicals that have worked in the past – in precise quantities until one has the desired effect. If a scientist is lucky enough for crystals to form from proteins, they still have to fish out the tiny micrometre- sized crystals using miniature loops of nylon. The work is very manual, requiring the utmost patience. Once the crystals are ready to be studied, research groups usually bring their whole team along: they will run shifts around the clock to make the most of their short beam time. During the pandemic, however, many of the research teams have been forced to work remotely while Campbell and her colleagues manage the on-site sample setup.

Campbell knows what it's like to have to run an experiment at such a facility remotely. During experiments for her PhD at the University of Cambridge, she would sit at a computer in her lab while her carefully prepared crystal samples were placed into the beam remotely by someone else at the UK synchrotron light source, which is called Diamond. She'd click refresh and a new shape of a protein structure would appear on her screen. While she was gaining insights into proteins, the actual geometry of the whole experiment was obscured. Now she was on the other side of the equation, helping remote users run experiments to learn as much about the COVID virus as possible.

For the biologists Campbell works with, it's worth the remote setup and late nights. Without a synchrotron, they'd have to spend days on end using a lab-based X-ray source, taking about forty minutes for every image angle (crystallography involves taking images through 180 degrees, gathering the diffraction patterns and reconstructing the 3D structure through mathematics). At the MX-2 beamline Campbell runs, an experiment to collect 180 degrees of data takes just eighteen seconds. So if someone is trying to test a range of samples, for instance fifty slight variations of a protein, what would once take an entire PhD can now be done in a few hours of beamtime. In many cases, the unique properties of synchrotron light mean experiments can be run that simply weren't possible before. Without synchrotrons, biologists would have taken years to understand the structures of SARS-COV-2.

Around the world at facilities like this, scientists pulled together towards one overriding goal: they were looking to make atomic-scale maps of as many of the proteins that make up SARS-COV-2 as possible. In less hurried times, researchers used facilities like this to create images and unravel structures of many key biological molecules, leading to new treatments for AIDS, skin cancer, type 2 diabetes, leukaemia and seasonal flu, as well as breakthroughs against ebola, zika and SARS viruses. This is why the roughly fifty synchrotron light sources around the world form part of our front-line defence against emerging viral diseases.

*

When the first of those dedicated synchrotron light sources, the Daresbury SRS, closed down in 2008, a study was conducted of the 11,000 scientific users over its lifetime. It identified thousands of discoveries that have affected our lives in direct and indirect ways. New materials for clothing and electronics, new pharmaceuticals and new detergents are just a few of the products that emerged from the studies at this one facility. It's hard to get an idea of just how far-reaching the uses of such a facility are, but when it closed, eleven out of the top twenty-five UK companies ranked by research and development (R&D) had used it.

The SRS was used to figure out the structure of foot-and-mouth disease, leading to new vaccines, and to understanding a phenomenon called 'giant magnetoresistance' or GMR, which is the trick behind the immense storage capacity in our electronic devices like iPhones. SRS research contributed to cleaner fuel and a range of new medicines. It even contributed to cultural heritage, studying samples from the Tudor warship the *Mary Rose* to learn how to better conserve the remains. A study performed by Cadbury – the chocolate makers – used the light to study the formation of crystals in chocolate, making the 'mouthfeel' of the chocolate even more delicious. A similar technique was used to study crystal formation in metals to improve aircraft safety.

Newsworthy breakthroughs are the bread and butter of these facilities. They churn out science at a rate that can be hard to keep up with. The story of synchrotron light forces us to appreciate the extent to which the tools of physics can transform other areas of science. It reminds us that different areas of knowledge are inseparable, from the very smallest to the very largest objects in nature and everything in between. As Campbell says, it makes her feel small just walking into this big facility every day: sometimes it strikes her as quite overwhelming to consider just how complicated synchrotrons are. The accelerator physics team would surely say the same of her work. That's why many modern scientific breakthroughs are necessarily interdisciplinary: no single individual can fully understand the entire facility. Yet using this product of physics research, scientists like Campbell and her predecessors can create knowledge that is much further-reaching than the General Electric physicists, Lawrence, Kerst or Oliphant could ever have predicted.

As we've seen, this knowledge extends beyond biology and even beyond our Earth. Understanding the basic science behind synchrotron radiation helped open up a great tool for astronomy. It let astronomers see the objects in space in a totally new light, revealing the inner workings of everything from galaxies to quasars and black holes, as all of them are emitting synchrotron light in the form of radio waves. Today, radio astronomers are studying the complex behaviour of magnetic fields generated in extreme regions of the Universe, like recent observations of 'fast radio bursts': extremely powerful millisecond-length pulses of radio waves which point to new, high-energy processes, which we do not yet fully understand. Cosmologists, meanwhile, are looking at the existence of magnetism in remote regions of the cosmos as an explanation for the rapid inflation of the early Universe. In synchrotron radiation, physicists have a tool that unites them in their quest to understand the physics of the very large and the very small.

This is all possible because the principles of physics apply not just on Earth, but as far as we know, everywhere. The same physics that allows us to uncover the mysteries of the outer reaches of the Universe lets us unravel the inner workings of our biology and to intervene when it goes wrong. There's no particular reason the Universe should work this way, but it is deeply and profoundly fascinating that it does.

In the end, synchrotron radiation, which proved such an incredible tool for astronomers and other scientists, became a huge barrier to particle physicists. They wanted to push particles to higher and higher energies in order to smash atoms, but now they were faced with the fact that particles would radiate away energy when they tried to push them faster. They would have to pump in even more power to overcome the loss of energy as the particles sped around the machine. Before long they would reach the practical limit of how much energy they could give to particles – at least, for some types of particles.

The radiation formula predicted that accelerating low-mass particles like electrons to high energy was going to be a problem, but that the emitted radiation power would be far lower for heavier particles. A proton is almost 2,000 times heavier than an electron but would emit an astonishing 10^{13} times less radiation than electrons.[17] The flip side is the challenge involved in bending the high-energy protons in a circular accelerator, which required either very strong magnets or a much larger ring than the room-sized electron accelerators. As physicists were determined to push protons to higher energies, this limitation meant that one thing was inevitable: the particle accelerators built in the second half of the twentieth century would grow and grow.

Physicists had to join forces and assemble specialised teams of engineers, data analysts, managers and others in order to build and operate enormous machines. They became some of the earliest adopters of computing technology, and had to create new ways of seeing particles, always pushing the limits of the possible. In time, their search would reveal far more particles than anyone guessed existed. Hundreds of researchers sought to answer the question: is there an underlying order in nature? Can we predict and classify the many different particles, or is our reality simply a manageable kind of chaos?

Part 3

The Standard Model and Beyond

The truth is not for all men but only for those who seek it.
 – Ayn Rand, *Anthem*, 1938

8

Particle Physics Goes Large: The Strange Resonances

Luis Alvarez was falling asleep as the plane he flew in, *Great Artiste*, approached Japan. It was just before dawn on 6 August 1945 and the thirty-four-year-old physicist was exhausted. His pilot was following another aircraft, the B-29 bomber *Enola Gay*. A third unnamed plane, later dubbed *Necessary Evil*, flew nearby. Unlike most Second World War bombing missions which involved hundreds of aircraft, this one involved just three: they would fly in stealthily and drop a single bomb on the city of Hiroshima. It was no ordinary weapon: it was Little Boy, an atomic bomb filled with enriched uranium.

Alvarez had used his physics skills to help develop Little Boy as part of the Manhattan Project, the highly secretive US-based project, conceived with UK and Canadian allies, which developed the first nuclear weapons. Throughout the war the project grew to an enormous enterprise that employed 100,000 people, most of whom were entirely unaware of the goal of their work. When the decision was made by military leaders to use the new weapons on Japan, Alvarez[1] was tasked with deploying instruments that could follow the bomb descent and record measurements of the energy released when it exploded. While his equipment wore a parachute, Alvarez chose not to: if they were shot down, he reasoned it was better to die than be captured by the Japanese.

When the moment came, the bomb was released from the bay and dropped for forty-four seconds before being detonated. A small internal explosion brought two pieces of highly enriched uranium together to form a critical mass. Uranium-235 nuclei then fissioned apart, releasing neutrons and creating more fission. A chain reaction ensued. A blinding pulse of light filled Alvarez's aircraft, followed by a series of shock waves that threatened to break apart the plane. It took a full ten minutes for the mushroom cloud to rise to 60,000 feet. After that, Alvarez peered down to look at the landscape beneath. It was barren. He later wrote that he 'looked in vain for the city that had been our target', thinking perhaps they had missed. The pilot had to set him straight: the target city of Hiroshima had been destroyed. Eighty thousand people were killed in an instant.

On the flight back to base, as the momentous nature of the mission sank in, Alvarez wrote a letter to his four-year-old son. He knew that it would be hard for his son to understand how he could have been involved in this particular historical event. The Alvarez family had plenty of adventures behind them: Alvarez's grandfather had run off to Cuba, studied medicine in California, then married his grandmother (who grew up on a mission in China) and moved the family to Hawaii. His father (also a physician) and mother had spent time working in Mexico before returning to San Francisco, where Alvarez was born. Tall, blond, courageous and intelligent, Alvarez chose physics as he sensed it would lead to adventure. But war work was not the type of adventure he'd originally had in mind.

Three days later Alvarez watched from the island of Tinian as his colleagues flew off to accompany a second bomb, which was dropped on the city of Nagasaki. One day after that, on 10 August 1945, the Japanese offered to surrender. Alvarez did not write about the events again for another forty years.

Today the Peace Memorial Museum in Hiroshima tells the story of the devastating impact that nuclear weapons had on the city, and explores the wider ramifications of their use in the Second World War. For a visiting physicist it feels particularly disquieting: a startling number of famous names in our field appear in the museum's description of the Manhattan Project. Many of the characters we've met in this book so far were involved in the development of nuclear weapons, because they were the ones with the knowledge and skills required for the project. Ernest Lawrence's cyclotrons were converted to separate uranium isotopes, and he oversaw a vast project building calutrons (isotope separation devices) based on the expertise his team had gained constructing machines at Berkeley. A number of Lawrence's staff and students were involved, including Alvarez. Seth Neddermeyer came up with the idea of an implosion to bring together the critical mass for the plutonium bomb that was dropped on Nagasaki. Niels Bohr, James Chadwick, John Cockcroft and Mark Oliphant were all part of the project, alongside many theorists who have played a less prominent role in our story, including Lawrence's colleague Robert Oppenheimer, who famously ended up leading the project.

Some physicists were invited but turned down roles at the Manhattan Project to do other work during the war. Carl Anderson was asked to be head of the project but, as he needed to support and care for his mother,[2] he worked on artillery rockets instead. One physicist who outright refused to participate was Lise Meitner, one of the only women in the field at the time. Nicknamed 'the German Marie Curie' by Einstein, Meitner was originally from Vienna. She had to study physics privately, because public universities wouldn't admit women. Encouraged and financially supported by her father, she went to Berlin after obtaining her PhD. There she somehow convinced Max Planck to allow her to take his lectures, and eventually became his assistant. Later, after becoming the first female professor of physics in Germany, she had to flee the country owing to her Jewish heritage. Lise Meitner was the person who first realised that rather than just emitting beta or alpha particles, nuclei could split apart entirely, and her nephew Otto Frisch coined the term 'nuclear fission' to describe this idea.[3] Despite the potential to use her expertise, she refused to join the Manhattan Project, saying, 'I will have nothing to do with a bomb!' Meitner's colleague Otto Hahn had published the first evidence for nuclear fission without listing her as co-author, to avoid giving away the fact of his correspondence with her and finding himself persecuted as a result. Hahn was awarded the Nobel Prize for the work in 1944. Meitner's contribution was not recognised.

For those who did agree to join the Manhattan Project, they did not know if the problem they'd been tasked with solving – to make a nuclear weapon – was even tractable. But after watching the first explosion in July 1945, called the 'Trinity' test, it became clear that it was indeed possible. That reality horrified many of the physicists, who created petitions in Chicago and at Los Alamos objecting to the deployment of the weapon they had created. But the decision was not in their control. After the bombs were dropped and the cities of Hiroshima and Nagasaki destroyed, the mood among the physicists at Los Alamos was sombre. As Evelyne Litz, who worked in health physics and as a librarian, later recalled, 'The day the bomb was dropped there was no hilarity […] None of our friends got together; we were very solemn.'[4] Many physicists like Alvarez would not speak about the event for a long time. Most would later summon the matter-of-fact narrative that it had helped end the war and thus saved lives on both sides overall. Whatever their moral position, for the vast majority of them, the work was done.

Physicists emerged from the Second World War less naive and more socially conscious than they had been before. They were not quite seeking atonement, but in the post-war era there was definitely a renewed dedication towards using their skills for peaceful societal benefit. The war had seen physics used for destructive purposes, but now the time was ripe for a more noble pursuit: the creation of knowledge and the discovery of new particles. Like the Manhattan Project, this challenging endeavour would require broad collaboration, which the United States had the proven capacity to deliver. Physicists began to adopt a new large-scale approach to their work, and it would pay off for science and for society.

On 16 August 1945, Winston Churchill declared that 'America at this moment stands at the summit of the world'. He was sharing with the House of Commons his desire to keep the secrets of atomic weapons confidential, for 'the common safety of the world'. The United States had created immense military-industrial capabilities, which Churchill believed gave the country a new post-war obligation. He continued: 'Let them act up to the level of their responsibilities, not for themselves but for others, for all men in all lands, and then a brighter day may dawn upon human history.'[5]

For many young physicists like Alvarez their research work had been completely disrupted by the war. Now each faced a choice: what to do next? Most of the physicists went back to their universities and research labs. Alvarez returned to Berkeley determined to put his radar knowledge to use in particle accelerators.

His choice was driven by the knowledge that he'd be working on the best machine in the world. With the US government's financial backing, the Berkeley team completed the large cyclotron they had been building prior to the war, with one change: they incorporated the idea of phase stability from Edwin McMillan[6] (see Chapter 7) and instead built a proton 'synchrocyclotron' that reached an unprecedented 350MeV beam energy. The Berkeley team set to work finding new particles.

First they used the accelerator to replicate discoveries made using cosmic rays. Mountain-top experiments with cloud chambers and nuclear emulsions had been a productive method to discover positrons, muons and pions as we saw in Chapter 4. Now evidence was emerging of new particles that had very different properties from the ones they'd seen before – such as electrically neutral 'V' particles (1947), identified by their decay into pairs of tracks that formed a V-shape in the detectors. In 1949 another particle was found that decayed into three pions,[7] later called the kaon, and in 1952 a new particle called a Xi-minus hyperon ('hyper' because it was heavier than a proton) was found in cosmic rays.[8]

Nature seemed to be teeming with particles that played no role in everyday matter, and it wasn't clear what their significance was. To make matters worse, most of the new particles seemed to live far longer than expected – where long means nanoseconds – which left theorists scratching their heads. The new particles became known as 'strange' particles. With only a few photographs of each, there was nowhere near enough data coming from cosmic rays to understand the new particles fully. The only way to make sense of these mysterious particles was to create large quantities of them in the lab.

Berkeley's new large cyclotron provided a turning point. In 1949 colleagues working with Alvarez and Lawrence's 350MeV accelerator found a particle that high-altitude cloud chambers had missed: an electrically neutral version of the pion.[9] This event marked the first time an unknown particle had been found using an accelerator rather than cosmic rays. Finally, accelerator technology was reaching unprecedented energies and with far more mature, reliable machines, physicists could start to move beyond what could be achieved with cosmic ray experiments. Particle accelerators provided the controlled conditions needed to piece together the complex puzzle of particles and forces. The only problem was that 350MeV wasn't a high enough energy to really understand the full picture.

The energy reach of the accelerator was crucial because strange particles are *heavy* – that is, they have a higher mass than the earlier discoveries like the muon and pion. The equivalence between energy and mass is given by Einstein's $E=mc^2$, and is so ingrained for particle physicists that we even use units of energy to describe particle masses. The neutral pion (π^0), for example, has a mass of 135MeV, which is its *rest mass* – the mass measured when stationary – but given in units of energy (MeV). This equivalence between particle masses and energies means that $E=mc^2$ gives us the exchange rate between mass and energy. It's an absolutely staggering exchange rate because c, the speed of light, is 299,792,458 metres per second. Squared, it's a number so large I wouldn't dare write it down on the page. This was no longer just a theoretical exchange rate: with large accelerators it was now an experimental reality.

Building accelerators to reach higher energies was no longer just to explore the neutrons and protons in the nucleus. What scientists wanted, although they wouldn't have quite phrased it this way at the time, was to create entirely new particles out of the vacuum, from energy. This can be a little difficult to get your head around at first. The basic principle is that we bring in high-energy particles – in this case protons – and smash them into a target. The original particles disappear and all that energy is converted into *new* particles, new matter. The original particle simply ceases to exist – which is counter-intuitive in the classical realm, but allowed in quantum mechanics.

There are of course rules about such things: nature doesn't let you smash any particle into any target and produce whatever you like. Certain quantities have to be conserved. For example, the total energy of the particles has to be the same coming into the collision as it is going out. When you smash a beam of particles into a target, much of this energy doesn't go into making new particles, but instead gets carried away as kinetic energy in the debris. There are other rules governing these particle interactions, including the conservation of *electric charge*, *angular momentum* (a particle can spin on its axis) and other quantum numbers, but more on that later. For now, what matters is that, to create strange particles, the Berkeley physicists knew that they needed a higher energy proton beam than a cyclotron could ever provide.

For Alvarez and Lawrence a big new goal came into view: building a machine that was powerful enough to create all the known strange particles found in cosmic rays, and perhaps even heavier ones. To achieve that, they had to build a new type of machine. Rather than a cyclotron, which required one enormous magnet – the magnet for the 350MeV cyclotron was so large the team of a hundred people were photographed sitting in the iron yoke – they would build an accelerator composed of a ring of many smaller magnets. The Berkeley team began making plans for a *proton synchrotron*[10] – a ring-like machine distinct from the synchrocyclotron that had come before – that could reach the same energies as particles coming from cosmic rays. Since it would reach up to billions of electron volts, the 'BeV' range, the machine's name was easy: they called it the 'Bevatron'.

The Berkeley team were not alone in their ambitions. Across the country on Long Island, eleven different universities had collaborated to form the new Brookhaven National Laboratory and the construction of their own large proton synchrotron was already well underway. In 1953 they switched on the 'Cosmotron' – a 23-metre copper-coloured ring made up of 288 magnets each weighing six tons. It was a thing of industrial-scale beauty. Inside all that copper and iron was a beam pipe where protons would be accelerated to 88 per cent of the speed of light. When the Cosmotron reached its design energy of 3.3GeV (the name BeV didn't last and we now use 'giga' for a billion, hence 'GeV') it took the record as the highest energy accelerator in the world, out-performing the cyclotron at Berkeley by almost a factor of ten.

The Berkeley team pushed ahead and in 1954, just a year after the Cosmotron was turned on, the Bevatron roared to life. There was no mistaking when it was operational: an enormous motor-generator ramped up and down, filling the concrete hall with wailing sounds. The Bevatron was even larger than the Cosmotron, stretching 41 metres across with an evacuated beam pipe so large it was said you could almost drive a car through it. Alvarez and his colleagues – chiefly physicist Ed Lofgren and engineer William Brobeck – had outdone their rivals to reach almost twice the energy of the Cosmotron, producing a proton beam at a world record 6.2GeV.

Why were two accelerators built instead of one? Aside from the geographical distance between the two labs and the concentrated research communities on the East and West Coast, the reason is primarily that the US government had made the decision to continue operating the large laboratories it had created during the war, pooling resources of people and finance towards large scientific goals. They also endorsed new labs like Brookhaven because they believed that having multiple labs enabled a level of spirited competition.

The technological developments of the Second World War demonstrated that a team of physicists and engineers with sufficient resources could solve incredibly difficult theoretical and practical problems. What's more, they showed an ability to work in teams of unprecedented size and complexity, with hundreds of scientists and engineers working alongside tens of thousands of other staff from construction workers to fire crew, towards some of the most challenging goals humans had ever conceived of. Their way of working became the prototype for other ambitious scientific projects, including the US space programme (and that of the Soviet Union). From this time on, physics, particularly in the United States, was afforded a status that other subjects weren't.

This new support for physics coincided with an enormous growth period in the United States. The economy boomed, bringing new consumer goods, new suburbs and new wealth. The birth rate swelled, with a record 3.4 million babies born in 1946 alone. The government budget expanded too, with investments in interstate highways, schools, military operations and new technologies like computers. As a result, in the 1950s and 1960s particle physics boomed as well. The physicists themselves gained a new sense of confidence. Big questions were now within their grasp: what are the strange particles found in cosmic rays, and what could they learn from them – about the Universe, about matter and about the forces that bind everything together? Do all the particles they had discovered have an antimatter equivalent? Is there an underlying order to everything?

Experiments outgrew university labs and became national facilities, uniting large groups of people in pursuit of a common goal. Alvarez and Lawrence were just two of the many physicists involved in this change. The experiments built in this period were focused on large particle accelerators, starting with the Bevatron and Cosmotron, which would eventually feed particles into new detectors that pumped out millions of images that needed analysing. Even the meaning of the word 'experiment' in the physics lexicon started to shift.

As we have seen, in the early days a researcher would build their own equipment from scratch, or at the very least operate it themselves. They would design an experiment as a test or trial of an idea. By the 1950s an experiment came to entail a gigantic piece of machinery designed by one group, maintained by specialist engineers, operated by dedicated staff, the results of which were analysed by one team and interpreted by yet another. Multiple groups could use the same experiment to look for different things, and the accelerators, detectors and other parts of the whole would be tweaked and upgraded as new technologies were invented and adopted. It became hard to say where one experiment ended and another one started.

Today researchers in particle physics are used to large laboratories and international collaborations, but these weren't always the norm. It was in the mid-twentieth century that the technological, political, scientific and personal coalesced into the age of Big Science which gave us our modern way of doing particle physics. As a result, the number of particles being discovered exploded, with experiments getting so far ahead of theory that it took almost twenty years for the underlying order to be understood in a mathematical sense.

Sitting in front of control panels filled with dials and meters, dedicated accelerator operators would coax the Cosmotron (on the East Coast) and the Bevatron (on the West Coast) up to full energy and then direct the beam onto a target, creating a prodigious source of rare particles. Before long the teams had produced and measured all the known cosmic ray particles: pions, muons, positrons and strange particles. Now, instead of observing individual pion events in a painstaking analysis of cosmic rays, accelerators could produce a steady beam of pions with large amounts of energy, ready to be analysed in detail. At the Cosmotron in 1953 pions were fired into a cloud chamber and produced a huge number of strange particles on demand, and the Bevatron soon followed suit. With accelerators, physicists had a data rate that the cosmic ray pioneers could have only dreamed of.

By 1954, when the Bevatron started operating, the list of strange particles had grown: by Alvarez's account there were 'several charged particles and a neutral one all with masses in the neighborhood of 500MeV'[11] as well as three heavier than the proton, the neutral lambda (Λ), the two charged sigmas ($\Sigma^\pm$) and the negative xi (Ξ^-). Far from suddenly answering all their questions, the list of oddities only increased as they made more measurements. The strange particles continued to emerge with lifetimes 100 billion times longer than expected. Not that they lived long in objective terms – they only lasted 10^{-10} seconds before decaying, a million times faster than the blink of an eye – but theorists' best efforts predicted they would decay in just 10^{-21} seconds, a hundred billion times shorter again! Furthermore, some particles they expected to be produced in equal quantities were not.

At this point physicists believed that there were four forces in nature. Gravity and electromagnetism were well known, but couldn't explain the nuclear realm, so two others were proposed. The *strong* nuclear force was put forward by Hideki Yukawa in 1934 as the force which binds the protons and neutrons together in the nucleus. His theory came with a particle with a mass around 200 times that of the electron, which would carry or *mediate* this force. At first the muon was thought to be the carrier of the strong force, but it was soon ruled out as it didn't interact with nuclear matter in the expected way. Later the pion seemed more likely, but this still wasn't quite clear. The second proposed force was the *weak* nuclear force responsible for radioactive beta decay, described in a theory by Enrico Fermi back in 1933. Where exactly the strange particles fitted in this picture was unknown. Could it be that the strange particles were produced through one force, the strong nuclear force, but decayed by another route via the weak force?

At the University of Michigan, a twenty-five-year-old experimental physicist named Donald Glaser had the problem of strange particles on his mind. Even in 1950 he could see that strange particles had led to the field of particle physics becoming, as he said, 'sort of stuck'.[12] Everybody in the field at the time knew why that was: they weren't collecting enough data. Without more data, the theorists didn't have enough information to figure out what strange particles were, or how they fitted together with other particles and forces in nature. Glaser set out to come up with a method to change all that.

Building big accelerators alone would not solve all the mysteries of strange particles. Sure, an accelerator might *create* more strange particles, but that was no good if you couldn't detect and measure them. As Alvarez and others got on with building big accelerators, Glaser's idea was to build a detector which could capture more data from cosmic rays than the existing cloud chamber.

Contrary to many other physicists at the time, Glaser was not interested in being part of the big laboratories, preferring instead to work with a small university-based group. He had carefully considered what kind of life he wanted to live. As an athletic person, he longed to live atop a mountain at a ski resort, skiing during the day while his experiment gathered data. In the evenings he would sift through and make discoveries of new particles. He knew of some Swiss researchers doing almost exactly that, creating a slow but steady trickle of new insights with plenty of time to contemplate.

When thinking about a new detector, Glaser knew that he would need to find a way of giving minuscule particle interactions a huge amplification to make them recordable. What he was after was a 'meta-stable' state, where a tiny amount of energy triggers a much larger effect, just as the cloud chamber uses the meta-stable state of a supersaturated vapour to trigger cloud droplet formation. At first he considered clouds, but when he found out a group at Brookhaven was trying to build a high-pressure cloud chamber that took twenty minutes to reset between pictures he decided it was no good, as it would never collect enough data. Glaser set out on a hunt for a new way of seeing particles.

Glaser imagined discovering a liquid which solidified when a particle went through it, forming something like a 'plastic Christmas tree' of particle decays and interactions. He dreamed he'd be able to fish out the plastic trees, measure all the angles and discover new particles that way. But when he tried it with a chemical solution, instead of getting Christmas trees, the mixture just turned to a sludgy brown. He didn't bother publishing it and moved on to the next idea. He tried using ice crystals in water, but realised it would take too long to melt the ice and reset. Glaser tried every physical, electrical and chemical setup he could imagine, but none of them seemed able to produce a usable record of particle events suitable for collecting data.

Then, one day in 1951, a thought about pressure cookers changed his fortunes. In a pressure cooker, water can be heated to above boiling point (100 degrees Celsius) before bubbles are produced. He asked himself: 'Could I put a liquid in a pressure cooker and get it sufficiently above its boiling point that if I pulled the lid off very quickly, before the explosion occurs, it would be unstable enough that it will be sensitive to a particle?'[13]

He tried a number of different liquids, trying to see whether they would form bubbles when exposed to a radiation source. Soda water had too large a surface tension to work, and ginger ale fared no better. At one point, he had the idea that a liquid with a small amount of alcohol might work, so he found a commonly available liquid that met the criteria: beer. The only problem was that alcohol was prohibited anywhere near the university, so he had to sneak a case of it into the department after hours. He put a bottle in a large beaker of hot oil, placed a cobalt-60 source – a powerful gamma-ray emitter – next to it, and pulled off the cap, waiting to see if the beer foamed differently because of the radiation source. The beer, he concluded, didn't seem to be affected by the cobalt source, but he'd forgotten to take into account another aspect of his late-night experiment. The hot beer foamed so rapidly as it heated that it exploded into the air, hitting the ceiling. The next morning Glaser was put in the unfortunate position of having to explain why the whole department stank of beer. His department chair – a teetotaller – was furious.[14]

Eventually Glaser studied the relevant chemical tables and came across a liquid called diethyl ether, which is usually used as an anaesthetic. Glaser created a small glass bulb, around the size of his thumb, and poured the diethyl ether into it. At about 3 a.m. one morning he superheated the ether using hot oil. Next, he took the cobalt-60 source and suddenly brought it near the bulb. The liquid exploded with bubbles. He did it again, and the same thing happened. He quickly added to the setup with a high-frame-rate camera and flash bulb from local engineering colleagues and managed to capture a picture of the gamma rays traversing the tiny detector. He had done it. Glaser had invented a new type of particle detector: the *bubble chamber*.[15]

Glaser realised his new invention should allow him to collect data at an enormous rate. In the bubble chamber, the liquid is a thousand times denser than air, so the chance of seeing a particle traverse the chamber is a thousand times higher than in a cloud chamber. He prepared a paper and arranged to present his work at the American Physical Society meeting in Washington in April 1953.

Arriving at the conference Glaser was distraught to find he was scheduled to speak on the last day: the day when all the older and more established physicists would have already left to catch their aeroplanes home. One evening he lamented his predicament over drinks with a group of old physicists, among them Luis Alvarez. Alvarez conceded that he too would have left the conference by then, but was curious about what Glaser was working on. When he learned about the bubble chamber, Alvarez immediately realised the implications of the younger man's idea: 'I had been unsuccessfully racking my brain to find an appropriate detector for the Bevatron, which was about to be turned on. It was immediately clear to me that Glaser's chamber filled the bill exactly.'[16]

Alvarez made sure two members of his team stayed behind to hear Glaser's talk. Both Alvarez and Glaser knew what had to happen for the bubble chamber to be useful for detecting particles at Bevatron. First, an obvious improvement would be to replace diethyl ether with liquid hydrogen – since hydrogen is mostly protons it would produce simple collisions of high-energy protons from the Bevatron with the protons in hydrogen. However, hydrogen is extremely explosive and liquid hydrogen is extremely cold – around -250 degrees Celsius – so it would have to be done very carefully. The second challenge was to scale the detector up to a large volume so that high-energy protons had enough space to interact in the hydrogen, create strange particles and leave long tracks that could be photographed and analysed.

Glaser, back in Michigan, knew he couldn't compete with the enormous resources and teams of engineers that Alvarez had at his disposal. He had dreamed of using the bubble chamber with high-energy cosmic rays from space while he lived an idyllic mountaintop lifestyle. Now he realised the problem with his dream: the bubble tracks appeared and disappeared so quickly that there was no reliable way of triggering the camera at the right time to photograph particle interactions from cosmic rays. By the time the electronics could open a camera shutter, the bubble track would be long gone. The only way he could use the bubble chamber was to combine it with one of the big accelerators, where the predictable timing of particles arriving would give him a chance of detecting interactions.

After all these years of actively distancing himself from working at one of the large labs, it seemed he had no choice. He gathered his students for a hard conversation and eventually, they all agreed that they would go to the large accelerators. Glaser built a propane-filled bubble chamber 15cm across and purchased a 40-foot trailer in which he and his graduate students loaded all the equipment and travelled across the country. First, he took his detector to use at the Cosmotron at Brookhaven. The first roll of film he used had just thirty-six pictures. Among the images were thirty to forty examples of rare decays, which had barely been glimpsed before from balloon flights and nuclear emulsions. When he came out of the darkroom, a huge crowd gathered. In his words, 'Well, I didn't know exactly what I was going to get, but I knew I was going to get things – I knew if it worked it was going to be big.'[17]

The bubble chamber had much faster cycle times and better resolution than a cloud chamber, keeping pace with the proliferation of particles produced by the new accelerators. Alvarez and his team had seen its potential and had immediately made plans to create a big hydrogen bubble chamber for the Bevatron. First, they re-created Glaser's results, then a small group from the mechanical workshop helped with building a series of hydrogen bubble chambers of increasing size. Glass bulbs were not strong enough, so they designed steel tanks fitted with glass windows through which the bubbles could be photographed.

By 1958 Alvarez had a 15-inch bubble chamber working at the Bevatron, and soon convinced Glaser to move to California with six or so graduate students. Alvarez was launching a programme to build an enormous 72-inch liquid hydrogen bubble chamber, but even before this, the smaller chambers produced reams of data. Soon the biggest challenge was keeping up with analysing the millions of photographs produced. Glaser's bubble chamber certainly solved the problem of having too little data, but created a new challenge: extracting useful data from the film required someone to look at every single picture one by one.

The photographs were sent to groups all over the world to study. Glaser had a special briefcase made with a film-viewer installed so he could analyse bubble chamber tracks on his many train travels between

Brookhaven, Michigan, Chicago and Berkeley. In time, this analysis became a specialised job undertaken by a group of trained 'scanners'. This was an almost entirely female group known as the 'scanning girls', who sat day after day analysing particle tracks.[18] At first they would measure the length and arc of interesting particle tracks, recording the data by hand, step by step. Alvarez's team eventually created semi-automated measuring machines that the scanners would use to put data onto punch cards and in early computers.

What emerged from this industrialised data-collecting was not the clear picture that was expected, but utter confusion. In 1958 Alvarez found a confounding new particle which was called a $Y^*(1385)$ – pronounced 'Y star thirteen eighty-five' – named for its mass of around 1385MeV. I say 'around' because the truth was that its mass was uncertain, and this was a key part of its mystery. In truth, the masses of all particles are uncertain; the precision with which we know the mass is related to how long they live. This is not, to be clear, because of a measurement error, but is a property of matter enshrined in a key principle of quantum mechanics: *Heisenberg's uncertainty principle*. This principle implies that the shorter the lifetime of a particle, the less certain we are of its energy – and thus its mass. Alvarez's new Y^* particle only lived around 10^{-23} seconds, which is why its mass was only 'about' 1385MeV. What Alvarez found was not just a new particle, but the most transient physical phenomenon in the natural world: even travelling nearly at the speed of light they move less than the width of a proton before decaying.

The $Y^*(1385)$ was the first of a completely new type of particle called a *resonance* particle, and many more followed. At the start-up of the Bevatron there were around thirty known particles but eventually around 200 new particles and resonances were discovered – so many that they more than used up the letters of the Greek alphabet. While the experimentalists were making discovery after discovery, the theoretical physicists were devising their own creative revolution to try to bring order to the new particles.

First, the strange particles led the way. In 1956 theoretical physicist Murray Gell-Mann[19] (and independently Kazuhiko Nishijima in 1953) had assigned each of the strange particles a new quantity called *strangeness*. The idea was that strangeness was conserved in *strong* interactions: if two particles were created and one had strangeness +1 and the other -1, overall strangeness was conserved. Having observed that strange particles were usually created in pairs, this seemed to fit. Gell-Mann also suggested a reason why strange particles seemed to live longer than expected: strangeness, he predicted, was *not* conserved in weak decays. When strange particles decay to non-strange ones in this way the decay can't proceed by the strong interaction (which must obey strangeness conservation) but rather must undergo a weak decay on a slower timescale. Their decay is inhibited by nature, which explains the relatively long lifetimes of strange particles.

By 1961 Murray Gell-Mann and Yuval Ne'eman had each proposed a classification system based on strangeness and electric charge, often referred to as the 'Eightfold way' – a riff on the Noble Eightfold Path of Buddhism. Using detailed mathematics to underpin their theory, Gell-Mann and Ne'eman were able to simplify the abundance of particles into orderly groups, creating a classification system. One aspect of the classification was differing particle *spin*, a quantum number describing the intrinsic angular momentum of a particle spinning on its own axis. The pions and kaons (all spin-0) form a group of eight *mesons* while the lambda, proton and neutron (spin-1/2) were part of another octet of so-called *baryons*. There was a different group of ten baryons – a decuplet (all spin-3/2) that included some of the odder particles like the delta, sigma and xi, all of which had been observed. Here was the crunch: in the decuplet, the theory made a prediction that there should be an as-yet-unseen particle, called the omega minus. There was only one way to confirm that Gell-Mann's system was correct.

By 1964 Brookhaven had upgraded the Cosmotron to a new accelerator called the AGS (Alternating Gradient Synchrotron)[20] and installed an enormous 80-inch liquid hydrogen bubble chamber with a 400-ton magnet. A team led by Nicholas Samios set about searching for the omega minus. When it was discovered later that year, it was a major triumph for the new theory. They were heading in the right direction.

After this discovery, the mathematical underpinnings of Gell-Mann's classification system led him to a truly astounding suggestion: that protons, neutrons, mesons (like pions) and resonances were not truly fundamental particles at all, but were instead composed of smaller pieces. Gell-Mann called these fundamental constituents 'quarks'.[21] He proposed there were three types of quark, known as 'up', 'down' and 'strange'. Up and down quarks made up the proton and neutron, while the strange quarks went into making the strange particles – kaons, lambdas and so on. The resonance particles could be understood as excited states of quarks combined together.

Industrial-scale Big Science was starting to pay off. There seems almost no way that we could have arrived at the idea of quarks without it: it is simply not possible to build and operate such massive equipment with a small team. Of course, such massive expansion comes with its problems: in retrospect, it can be difficult to even find out who exactly was involved in a discovery, or what their precise roles were. Very few first-hand records of the scanning girls exist. Graduate students who left the field don't get detailed biographies. The omega minus discovery paper had thirty-three authors and this didn't include any of the accelerator designers, engineers, scanners or theorists – not even Gell-Mann.[22] As a result, today we usually only hear the stories of the few theoretical physicists rather than the teams of experimentalists, engineers and others needed to actually make discoveries like resonance particles and the omega minus happen.

Glaser, ever the champion of small science, was perturbed by this enormous shift in working style. Within just a few years of being awarded the Nobel Prize in 1959 for the bubble chamber, Glaser was sick of the administrative work involved in overseeing large teams of scanners and engineers and quit physics to work in neurobiology, where he went on to found the first biotech company, Cetus Corporation.[23] Alvarez, meanwhile, won the Nobel Prize in 1968.

Big Science as practised at labs like Berkeley brings together different types of scientists, which creates the ability to undertake ambitious applied research as well as curiosity-driven physics. Alvarez became an advocate of this style of research, and another Manhattan Project veteran, Robert Rathbun ('Bob') Wilson, would become known for it also. Like Alvarez, Wilson was one of Ernest Lawrence's former cyclotroneers but instead of Berkeley he moved at first to Harvard after the war. Wilson wasn't proud of his role in the development of atomic weapons, commenting in an interview 'I always hoped we would not be successful'.[24] Wilson grew up in Wyoming and his forebears had been Quakers. Even before the war he had been a pacifist, but after his wartime experience, he came to feel strongly about contributing to peacetime applications of physics.

In 1946 Wilson came up with an idea which seemed to him so obvious he assumed it must have occurred to many others. The proton beam energies of cyclotrons were now high enough, a few hundred MeV, that the beams could reach deep into human tissue and might prove interesting for direct therapeutic use, particularly in cancer treatment. When his suggestion reached the medical community, it turned out that no one had really thought of doing this before. It would take many years, but his idea would eventually pave the way for the creation of a whole new type of cancer therapy using high-energy charged particles, called particle therapy.[25]

The question he needed to answer was: how do high-energy particles interact with the human body, and could that be used for cancer therapy? Being aware of the successful history of cyclotron-generated isotopes in medicine, he knew that Ernest Lawrence's brother, John, was the right person to approach to make progress.

In the 1950s cancer treatment using radiation was becoming mainstream. X-rays (and sometimes electrons) were being used for radiotherapy because it was well-established that *ionising radiation* – radiation with sufficient energy to dislodge electrons and form ions – can kill cancer cells. The aim with these kinds of treatments is always to deliver enough dose to the tumour to destroy it but as little dose as possible to the healthy tissue, which is difficult to achieve. It's hard because of the physics of how beams behave in matter, which was why Wilson's idea was a huge breakthrough.

When a high-energy photon or electron enters human tissue, which is around 70 per cent water, it interacts with the electrons around atoms and loses energy quite quickly. In terms of radiation dose that means it deposits a large amount just under the skin and a smaller amount deep inside the body. But when a heavy charged particle enters tissue or water, the tiny electrons aren't enough to slow it down, so it loses energy slowly, deviating only a little from its path. A proton or other heavy charged particle can get deep inside the body, depositing very little energy at first before it slows down and eventually comes to a halt, delivering most of its energy (and thus damage) at the end of its range. If you plot the depth versus dose of a proton, it follows a curve called a Bragg peak.[26]

Wilson realised that in terms of biology, the Bragg peak of heavy charged particles should be much better suited to treating tumours deep inside the body. By varying the starting energy of the protons, they could stop at different depths, allowing clinicians to pinpoint radiation where it was needed. But the physics of radiation and matter is one thing; the biological effect in a human still had to be figured out.

John Lawrence and his colleague Dr Robert Stone had previously investigated the use of neutrons for therapy, but their results were inconclusive. Wilson's idea of using charged particles instead led their colleague Cornelius Tobias to undertake biology experiments using the 350MeV cyclotron in 1948, testing out the effect of protons and deuterons on cells. When that looked promising, the first human was exposed to deuteron and helium ion beams in 1952. In 1954, the year they switched on the Bevatron, the first human was exposed to proton beams.

Despite having a sharper tool to deliver radiation deep into the body, the doctors couldn't use particle beams accurately because they would have been operating blindly: the imaging methods of the day didn't allow them to see deep inside the body because CT (Chapter 1) had not yet been invented. One target that *could* be seen was the pituitary gland, which controls the release of certain hormones. The first treatments therefore focused on stopping the pituitary gland from producing the hormones that cause cancer to grow. The very first patient, a woman with metastatic breast cancer, was successfully treated this way.[27] It would take decades until imaging and accelerator technologies coalesced to create a fully fledged cancer treatment, but this represented the start of one of medicine's most sophisticated techniques.

Today, more than a hundred centres around the world offer *particle therapy* – using either protons or heavy ions (usually carbon ions). This is up from twenty-two centres just a decade ago and the number continues to grow exponentially. Particle therapy is particularly well suited to deep and hard-to-reach tumours, difficult childhood cases or tumours near critical organs. In the UK a family made headlines in 2016 for absconding across Europe against the advice of their doctors for proton therapy for their child in Prague, at a time when the UK's first proton therapy centres were still being built.

Centres are carefully designed so that patients are barely aware of the fact that there is a particle accelerator nearby. At the Paul Scherrer Institute in Switzerland, patient treatment rooms are located down wood-panelled corridors with Japanese-style paper screens backlit to give the sense that daylight is just behind the wall. The screens mask the metre-thick concrete radiation shielding. During treatment, the patient lies on a carbon-fibre bed in the middle of a small room, mounted on a robotic positioning system. If it wasn't for a large white metal nozzle sticking out of the ceiling, you might expect a surgeon to pop in at any minute. But no human surgeons are required in this facility.

For curious patients (or physicists), a behind-the-scenes tour is allowed. The solid-looking walls of the patient treatment room reveal a hidden handle, which opens up into a cavernous space filled with big equipment, the sounds of vacuum pumps and buzzing power supplies. At the rear of the cavern a metal beam pipe emerges through a hole in concrete shielding, bringing protons in from the nearby accelerator. The proton beam travels through a series of magnets which climb up above the patient room. Finally, it is bent down through a 200-ton magnet that directs the beam to where it's needed. Yes, one of the magnets is almost twice as heavy as a blue whale, and yes, it often sits right above the patient.

This entire structure – called a gantry – is movable: it rotates all the way around the patient, delivering the beam from any angle while the patient lies on the treatment bed, unable to sense or feel the particle beam interacting with their body. For proton therapy a cyclotron a few metres across is just one small part of the full system. For heavier particles, synchrotrons around 20 metres in diameter are required.

The same physicists who design particle accelerators for particle physics are the ones who design synchrotrons (and some cyclotrons) for particle therapy in hospitals. The co-development of cancer therapy and particle physics was made possible by interdisciplinary collaboration. This was no mistake: it was the intention of Lawrence (who died in 1958) and his successors to create environments where knowledge could cross boundaries easily. This new large-scale team-based approach to science simultaneously revolutionised our understanding of particle physics and catalysed its benefits for society.

The drive now is to try to make this technology ever smaller, cheaper and more precise. Particle therapy is providing a completely new driving force for physicists to adapt and invent new accelerator technologies. It is just one of the many amazing practical applications of physics that have come out of the field's transition to huge collaborative experiments.

This transition happened not only in the United States, but across the world. As recovery from the Second World War got underway in Europe, French physicist Louis de Broglie first proposed that European scientists should band together to make a multi-national laboratory. It was a matter of necessity if they wanted to continue research in high-energy physics. They looked at the large projects being planned and built in the United States and knew that the only way they could stay in the game would be by pooling resources.

After lobbying governments for a number of years, twelve Western European countries ratified the creation in 1954 of a new laboratory, the Conseil Européen pour la Recherche Nucléaire, or CERN, near Geneva. It brought together researchers from countries which had been at war just a few years before, including Belgium, Denmark, France, Germany, Greece, Italy, the Netherlands, Norway, Sweden, Switzerland, the United Kingdom and Yugoslavia. Governed by a series of councils with representation from each member state, CERN created a unique structure for making decisions and pushing forward major scientific projects, encouraging nations to work together towards common goals. Unlike many of the US labs, it is written into CERN's convention that the laboratory 'shall have no concern with work for military requirements and the results of its experimental and theoretical work shall be published or otherwise made generally available.' The CERN remit was, and continues to be, science for peace.

Japan, meanwhile, had its scientific capability destroyed not just by war-induced poverty but by the actions of the US military in 1945. Afraid that cyclotrons might be used to develop nuclear weapons, occupying US army soldiers broke apart four of Japan's large cyclotrons and dumped them in Tokyo harbour.[28] It took until the 1952 Treaty of San Francisco, which restored peace between Japan and the Allied powers, for them to even be allowed to think of building new machines. Today, Japan has world-class research not just in particle physics but also in particle therapy.

For physicists in the 1960s, applications of their work to biology were still a side-project, the full realisation of which was far off in the future. With the new classification system of fundamental particles, they were finally coming to understand matter and forces on a more fundamental level. Not all new particles were elementary, with some having constituents called quarks, but quarks themselves had still not been observed. Particles containing quarks all interact through the strong nuclear force, but physicists, at this point, still hadn't solved the mystery of how the weak nuclear force worked. They only knew that this fourth force was responsible for beta decay. And it is beta decay that leads us to our next adventure.

9

Mega-detectors: Finding the Elusive Neutrino

Of the three basic types of radioactive decay – alpha, beta and gamma – one was oddly distinct from the others. Beta decay had troubled physicists since the early 1900s, as it seemed to violate one of the foundational laws of physics. The mystery of beta decay would take more than fifty years to resolve, forcing physicists to build a series of extraordinary underground experiments to hunt down a theoretical new particle which leading experts believed could never be detected. That particle was the *neutrino*: the most abundant, yet elusive, particle in the Universe.

From the early 1900s, experiments showed that beta radiation seemed to produce electrons with a range of different energies. This wasn't particularly worrying at the time, but after the atomic nucleus was revealed, problems began to surface. When an element undergoes beta decay, it is not left unchanged: the element shifts one place to the right in the periodic table. This is not the same as losing an electron from its atomic orbit, as this only changes the electric charge on the atom, not the *type* of atom. Beta decay, on the other hand, produces electrons from inside the nucleus. Detailed measurements by James Chadwick and colleagues showed that beta electrons could emerge with a *continuous spectrum* of energies ranging from very small up to some maximum energy, seemingly at random. This presented a profound challenge. Beta decay defied the most basic principles of physics.

In an atom undergoing beta decay there is at first one object, the atom. Afterwards, there are two objects, the atom and the electron. One of the key laws of physics, the law of conservation of momentum, dictates that the kinetic energy carried away by the projectiles in a simple two-body system like this should take a predictable, unique value. Alpha and gamma radiation obeyed this law nicely, but in beta radiation the energies were random and unpredictable. Breaking such a fundamental scientific principle is a sure sign that your experiment is flawed or your measurements are incorrect. Yet try as they might, anyone who did such an experiment couldn't get the data to come out any other way.

Every physicist had a different opinion on what was going on. Some, like Niels Bohr, contemplated throwing out the idea of momentum conservation, or at least sneaking around it by proposing that on the tiny scales inside atoms, energy might only be conserved on average, not in every single decay. One theorist in particular, Wolfgang Pauli, was unable to set the mystery aside. Pauli was well known for his critical and rational approach, which led to his nickname 'the scourge of God'. He wasn't happy with the suggestion of Dutch-American physicist Peter Debye, who told him at a meeting in Brussels to simply not think about beta decay at all. Pauli was determined to save momentum conservation and managed to come up with a theoretical solution, but to his horror it made the situation even worse. 'I have done a terrible thing,' he said. 'I have postulated a particle which cannot be detected.'

Pauli first presented his idea to other physicists in a letter in 1930. Perhaps, he suggested, a tiny electrically neutral particle was carrying away the energy? He felt it was so preposterous that he told his addressees he 'dare not publish anything' about it and that he was turning first to his 'dear radioactive people' to ask how likely it was to find experimental evidence for such a particle. The problem, as Pauli was well aware, was that these particles were predicted to have no mass and no electric charge, making it virtually impossible for them to show up in an experiment.

In 1933, Pauli's idea was turned into a fully fledged theory of beta decay by Enrico Fermi, an Italian physicist revered for both his theoretical and his experimental skills. Fermi dubbed the new particle the *neutrino*, or 'little neutral one', and submitted the theory to the journal *Nature*. It was rejected on the basis that it 'contained speculations too remote from reality to be of interest to the reader'. A year later in Manchester, physicists Rudolf Peierls and Hans Bethe calculated that the neutrinos created in beta decay could pass through the entire Earth without any interactions with matter. In fact, they could pass through quantities of lead with a thickness measured in light years. The neutrino might have solved the beta decay problem in theory, but what use is a particle if it is impossible to detect so it can't be verified? For years, it was more or less ignored by experimentalists.

The problem sat that way for two decades. Finally, in the 1950s, a thirty-three-year-old physicist decided to go after the elusive neutrino. That person was Fred Reines, who hailed from small-town New Jersey. He had barely finished his PhD work when he started on the Manhattan Project in the theory division, and he continued working in Los Alamos after the war. Reines's interest in physics had come in handy to his government, but like many of his peers, the war had redirected his expertise to the topic of atomic weapons. Reines decided it was time to do something of more fundamental importance to physics. After spending weeks in his office, the only idea that kept coming back to him was searching for neutrinos.

How could he make a source of neutrinos? How could he detect them? If he could build the right detector, perhaps he could prove their existence. A quick calculation told him that even if he could think of a way to build a detector, the likelihood of a neutrino interacting would be so low that the detector would have to be enormous. A liquid of some kind would work best, but the biggest liquid detectors in those days were about a litre in volume. (It was 1951 and Donald Glaser's bubble chamber was just emerging, although this couldn't detect the charge-neutral neutrino directly anyway.) How could he make a detector with a volume a thousand times larger than what was considered state of the art? Enrico Fermi had no idea how to make such a detector either.[1] If Fermi couldn't do it, how could anyone else? It seemed impossible, and for a while, Reines set the idea aside.

Not long after, he found himself stuck in the Kansas City airport after his plane was grounded due to engine trouble. A Los Alamos colleague, Clyde Cowan, was stuck there as well. Cowan was a chemical engineer and a former captain in the US Air Force who had worked on radar during the war. Where Reines was a sparkling extrovert, Cowan was more measured, less outgoing, but a brilliant experimentalist. The two wandered around the airport talking, and when Reines suggested his idea of searching for neutrinos, Cowan jumped at it. The pair decided they would go after the neutrino simply because everyone said it was impossible. Their managers at Los Alamos agreed to their outlandish proposal and, just like that, a new collaboration was born.

When they launched their project in 1951 a photograph of Reines and Cowan was taken with their core team of five, standing in a stairwell gathered around a cardboard sign. On it was a hand-drawn logo of a staring eye and the words 'Project Poltergeist'. Behind the sign, one of them was inexplicably holding a large broom in the air. They look in good spirits, as they'd need to be: their proposed experiment involved building an enormous tank, filling it with extremely well-filtered and prepared liquids, surrounding it in delicate electronics and hoping that they'd be able to catch a particle that was nigh-on invisible.

Reines and Cowan had studied Fermi's theory of neutrinos, which told them that neutrinos would only interact incredibly rarely, so they figured their first step was to find something that could supply as many neutrinos as possible. Although neutrinos can each travel a long way through matter, statistically if there were enough of them, one might by chance interact with a nucleus on its way through their detector. Their first idea was to try to catch the neutrinos from an atomic bomb, but they soon realised that the new technology of nuclear fission reactors offered them a less dangerous alternative. The nuclear reactor, they predicted, would produce an enormous flux of around ten-thousand billion (10^{13}) neutrinos per second per square centimetre. Not quite as many as a nuclear weapon, admittedly, but a steady source that could supply neutrinos over a very long time period.

Reines and Cowan focused their attention on searching for a reaction predicted by Fermi's theory in which a proton captures a neutrino, turning into a neutron and emitting a positron.[2] From this process, they expected to see a two-part signature from a neutrino. First, the positron would annihilate an electron and create a flash of gamma rays, which would be the tell-tale sign that a neutrino had visited the detector. The second part of the signature would come from the emerging neutron, which would be absorbed by a nucleus and emit a gamma ray roughly five microseconds later. What Project Poltergeist really needed was a system that could catch two gamma ray flashes, five microseconds apart. This signature, they hoped, would distinguish a neutrino from a cosmic ray or other background noise.

Having figured out what they were looking for, they designed a detector. Here, two recent technological advances came into play. The first was the discovery that some transparent organic liquids emit visible light when a gamma ray or charged particle passes through them. This 'liquid scintillator' would provide small flashes, which could then be picked up by another clever invention, the photomultiplier tube. These vacuum tubes physically look a bit like a long light bulb filled with electronics. When a flash of light hits the front of one of these vacuum tubes it is converted into electrons (via the photoelectric effect, as discussed in Chapter 3) which get amplified into an electrical pulse large enough for electronics to measure. The photomultiplier tubes would provide the eyes for the experiment.[3] As you can see, the team needed knowledge of chemistry and electronics as well as physics.

The team also took the step of designing a totally electronic method of measurement. There was no need to analyse millions of photographs, as in the cloud or bubble chamber. If neutrinos did interact in the liquid scintillator, the tubes would pick up the specific sequence of flashes and display them as blips on an oscilloscope.[4] The timing between the pulses would confirm the presence of a neutrino.

The downside to electronic measurement was that they were one step removed from what was going on in the experiment. It would be harder to intuitively understand the data when all they had to look at was a few blips. Any gamma ray flash in the detector opened the possibility that five microseconds later a coincidental flash might trick the physicists into thinking they had seen a neutrino. They had to make sure this didn't happen, but there was only one way of doing so: they had to remove every other possible source of radiation from the environment. Now the hard work began in earnest.

Reines and Cowan's working environment was a warehouse-style building, isolated and unheated. Trucks would continually arrive with parts for their experiment, filling up the building until they were surrounded by stacks of boxes twice their height. The team spent months testing different mixtures of the scintillator and measuring the response of the photomultiplier tubes to make sure the electronics worked. The lack of heating became its own challenge in winter, as the scintillator liquid had to be kept above 16 degrees Celsius to prevent it turning from clear to cloudy and ruining the experiment. They added electric heaters to make sure it could be kept warm, but couldn't afford the bill to keep themselves warm as well.

The first version of the detector was coming together in a prototype dubbed 'El Monstro'. When that showed them that the technologies ought to work together, they built a second detector, which they nicknamed 'Herr Auge' or 'Mr Eye'. Far from the litre-sized detectors that had existed before, they had now scaled up to a 300-litre capacity in a cylinder surrounded by ninety photomultiplier tubes.

Next they turned to the herculean task of eliminating sources of radiation that produced stray gamma rays in their detector. Some sources were obvious and could be predicted: the neutrons coming from the nuclear reactor could be blocked out with a thick layer of shielding made of blocks of paraffin wax. No money was wasted on ordering these from a specialist company. The team made the blocks themselves, clearing the snow outside their building and casting each block by hand, ready to be transported to the reactor site.

Other sources of radiation were more difficult to remove because Herr Auge was picking up radiation that their Geiger counters and other instruments didn't. Herr Auge turned out to be the best gamma ray detector ever built. It was so sensitive that they even lowered in team members to see if it could detect radiation from the human body. They found an easily detectable count rate coming from the small amount of radioactive potassium-40 in their secretary and colleagues.[5] The sensitivity turned out to be just what they needed, and it caused them to realise the detector could help build itself.

Before each new piece was constructed, they would place it in Herr Auge to measure the level of radioactivity. Brass and aluminium were more radioactive than iron and steel. Even the potassium in the glass of the photomultiplier tubes contributed to the background noise in the detector. Some radioactive components were found in the physical structure of the detector, which then had to be stripped apart and replaced. In every case they painstakingly swapped out any material producing background noise. It seems an extreme level of perfection, but they needed to be sure of the source of, quite literally, every flash of a photon in their detector, and it turned out there were a lot of sources.

After months of work they were ready to go. They transported and set up the detector near a nuclear reactor in Hanford, Washington. Then, they waited. They knew there would be no 'aha!' moment, just a gradual accumulation of individual events which they would analyse when they had collected enough. For a couple of months, the team took turns looking after the experiment, waiting and watching as their system sat silently in its heavily shielded enclosure.

When they regrouped and analysed the data, it looked like they had some light flashes corresponding to neutrinos. They were tantalising results, but not yet convincing. There was still too much noise in the data to be able to announce a discovery. The noise wasn't coming from human-made radiation or the materials of the detector, but from cosmic rays. They had worked so hard to reduce the stray radiation, but now they needed to eliminate this final source. There was only one feasible way to shield their experiment from the radiation coming from space: they would have to move it underground.

Luckily, a basement area was available over at the Savannah River Site, a nuclear reactor in South Carolina, and the owner was willing to let the physicists set up their experiment 12 metres beneath it. Reines and Cowan were now joined by several more Los Alamos colleagues to re-design and rebuild the whole detector.

By late 1955, Project Poltergeist was formally known as the Savannah River Neutrino Experiment. The setup had grown to a three-layered scintillating sandwich, with its rectangular tanks weighing in at a whopping 10 tons. The detector sat beneath the reactor, shrouded in its layers of shielding, while electronic cables carried signals to a trailer outside.

Reines and Cowan stayed at Savannah River running the experiment for about five months. Once all the chemistry and electronics were worked out, it came down simply to the careful collection of data, flash by flash. They were filled with hope each time they saw, just once or twice each hour, the characteristic blip-bloop of the two flashes five microseconds apart, which whispered *neutrino*.

They were determined to make sure it wasn't a fluke. Nothing was left to chance. They tested the detector with a positron source to make sure the light flash from the positron gave the correct 'blip' and then tried a neutron source to make sure that it gave the expected 'bloop'. They pumped the scintillating liquid all the way out, recalibrated the mixture to change the timing of the second light flash and checked that it had the desired effect. It did. All the while they recorded data, for 900 hours when the reactor was on and 250 hours when it was off.

In a final effort to be completely sure they weren't just seeing background neutrons from the reactor, they brought in trucks full of sandbags from the local sawmill and soaked them with water. One by one they hauled them to the experiment and built walls four feet thick around the detector. The mammoth effort provided enough additional water-shielding to block out any of the reactor neutrons. Yet there it was, blip-bloop. The neutrino signal stayed true.

Their eureka moment came not as a rush, but in a gradual accumulation of data until there were no doubts left. When all was added up, there were five times as many neutrino signals when the reactor was on compared to when it was off. From the 100 trillion (10^{14}) neutrinos that the reactor emitted each second, they had managed, against the odds, to design a system that could catch a few each hour and measure their interactions. Twenty-five years after Pauli predicted a particle that could not be detected, Reines and Cowan and their team had achieved the impossible.

'We are happy to inform you that we have definitely detected neutrinos' they wrote in a telegram to Pauli, who interrupted the meeting he was attending at CERN to read it out loud and deliver an impromptu mini-lecture. Legend has it Pauli later polished off an entire case of champagne with his friends, which might explain why his reply telegram never made it to Reines and Cowan. It read 'Everything comes to him who knows how to wait'.

The elusive neutrino had finally been found and the law of conservation of momentum was upheld even at the smallest scale, explaining the process of radioactive beta decay. The neutrino was not just a figment of theoretical imagining but a real, tangible thing in nature: an elusive, neutral, lightweight particle with the ability to travel to the deepest corners of the Universe without being stopped. The discovery of the neutrino opened up a whole new area of research.

From that first detection, more and more questions arose about neutrinos: what are their properties? Is there just one type, or more? Are they stable or do they have a limited lifetime? Which processes in the Universe create them? Like many of the experiments we've seen, Project Poltergeist created an avalanche of new questions, and over time most – but not all – of these questions have been answered. Ultimately, the elusive neutrino proved more important than previously thought. Neutrinos didn't just help us understand radioactive decay; they have led us to a new view on the Sun, supernovae and the origin of matter.

The increasing importance and richness of this field of research over time can be seen by the recognition of the Nobel committee. Three Nobel Prizes have been awarded for neutrino physics, all of them long after the original experiment. The first in 1995 went to Reines decades after their discovery (Cowan had sadly passed away thirteen years earlier), the second to Ray Davis and Masatoshi Koshiba in 2002, and the third to Takaaki Kajita and Arthur McDonald in 2015.

The original search for neutrinos was motivated by the mystery of beta decay, and Pauli's proposal of the neutrino came in 1933, just a year after Chadwick discovered the neutron. Now, we can bring these ideas together to better understand what happens in an atomic nucleus during beta decay: a neutron converts to a proton, changing the element type and releasing an electron (to balance the electric charge) and a neutrino.[6] The neutrino carries away some of the energy in the reaction, sharing the total available energy with the electron, which is why the electrons had unpredictable energies. Neither the electron nor the neutrino exist prior to the decay. The pieces of the puzzle started to fit together. But just as they did, a second competing experiment knocked physicists off their perch once again.

When the neutrino was first detected in the mid-1950s, it was only just dawning on physicists that the Sun was a nuclear furnace, producing its energy through a nuclear chain reaction called the 'p-p' chain, turning protons into helium through a number of steps.[7] If theories about the Sun were correct, a tremendous number of neutrinos should come flying straight out of the Sun at almost the speed of light, reaching the Earth roughly eight minutes later.[8]

Brookhaven radiochemist Ray Davis already had a head start a year before Reines and Cowan's first neutrino experiment. Davis wasn't looking for flashes of light. He was testing an idea posited by another theorist, Bruno Pontecorvo, who predicted that a neutrino interacting with an atom of chlorine would produce a radioactive atom of argon. Davis's expertise was in radiochemistry; if anyone was going to find a couple of individual radioactive argon atoms, he was the one.

Davis's attempt to detect neutrinos relied on using enormous vats of dry-cleaning fluid – a cheap and readily available liquid containing chlorine. He began with 3,800 litres and worked his way up from there. Despite his head start, Davis missed out on being first to discover neutrinos because nuclear reactors – and beta decay – actually produce their antimatter equivalent, antineutrinos, which is the type Cowan and Reines had detected.[9] Davis's experiment, however, was only capable of picking up the 'normal' kind of neutrino. Though he was beaten to the discovery by Cowan and Reines, in time Davis shifted his focus to detecting neutrinos not from reactors, but from the Sun. This decision was pivotal: it shifted neutrino physics from a curious side-effect of beta decay to the forefront of particle physics research.

Davis collaborated with a young theoretical physicist named John Bahcall, who did the difficult calculations to predict the rate of solar neutrino production. By 1964 the two collaborators published papers with their plans. They were confident that they could capture solar neutrinos, perhaps ten or twenty of them a week, but doing so would require an experiment a hundred times larger than their already enormous version – a prospect so ambitious that it made *Time* magazine before it had even been funded.

In 1965 an enormous cavern was excavated deep in the Homestake mine in South Dakota. In it they built a 380,000-litre tank and filled it with ten railroad cars full of dry-cleaning fluid. With incredible persistence and careful chemistry work, the mammoth effort paid off. By collecting a few dozen radioactive argon atoms, Davis managed to prove that he had detected solar neutrinos. The problem was that he only found around a third of the number of neutrinos that Bahcall had predicted.

They checked the calculations, but could find no errors. Davis went back to work and continued collecting data for almost another twenty years. All that time the mystery persisted: there was a strange lack of neutrinos coming from the Sun.

This solar neutrino problem raised the question: were the calculations wrong? Did they not understand how the Sun was generating energy? Was there something odd about neutrinos? Had the Sun stopped producing energy and were we – reliant on its output – in jeopardy? Eventually the favoured theory was that neutrinos were turning into something else, or disappearing between the Sun and the Earth. The idea of neutrinos behaving in this rather odd way had been suggested by Pontecorvo back in 1957,[10] but wasn't taken seriously for a long time. It was this question that motivated Art McDonald and around 100 other collaborators to build the Sudbury Neutrino Observatory (SNO).

McDonald, originally from Nova Scotia in Canada, had an early interest in mathematics and went on to study physics, obtaining his PhD in nuclear physics from Caltech in 1969. He left a professorship at Princeton to move back to Canada in 1989 to direct SNO. Under his leadership, SNO was built more than a mile underground in a nickel mine in Ontario, and working together with his 100 colleagues the enormous experiment ran between 1999 and 2006. Takaaki Kajita had led a similar experiment in a zinc mine in Japan called Super-Kamiokande. These two experiments would lead to a shared Nobel Prize in physics in 2015.

SNO is effectively an enormous underground cleanroom. Luckily, you can visit virtually, so can be spared the inconvenience[11] of a real-life visitor or scientist, who must shower, change clothes and then pass through air showers to keep any dirt from the mine out of the sensitive experiment at its heart. Once inside it feels quite austere. It's just the bare bones of a mine turned into a lab. The control room consists of five computer monitors on a few desks, sitting beside some racks full of equipment. Cable trays and pipes run down the wall above head-height. If it weren't for that rock you might forget that the experiment is almost 2,000 metres underground, the dangers of which are implied by a sign on the wall reminding the scientists, 'Safety and quality. Everywhere.' Visitors can walk – virtually – from the control room down a corridor and through a room full of machinery. Then they enter the detector vessel itself.

Virtually suspended within the empty detector, it feels like you've walked into an inside-out mirror ball. From every angle are 9,600 gold-hued photomultiplier tubes. Even through a computer screen the kaleidoscopic beauty of the 12-metre diameter geodesic sphere is breathtaking. A man clad in blue overalls and an orange hard-hat stands opposite, dwarfed by the experiment around him. The virtual tour was taken when the detector was empty, but normally all those golden detectors would be acting as the eyes of the experiment, looking inward to a thousand tons of heavy water loaned from Canada's reactor fleet worth an astonishing 300,000,000 Canadian dollars.

The wildest idea turned out to be correct. There are three types of neutrinos, and individual ones *oscillate*: that is, a neutrino born as one type, say as an electron neutrino, oscillates between its original state and the other two neutrino types, called muon neutrinos and tau neutrinos. Davis's experiment was only sensitive to electron neutrinos, so if solar neutrinos were oscillating to the other types, he would have missed out on detecting two-thirds of them. The first evidence for this came from Kajita's Japan-based detector Super-Kamiokande[12] in 1998, which consisted of 50,000 tons of ultrapure water in a tank 1,000 metres underground, with 13,000 photomultiplier tubes looking out for light flashes produced directly from neutrino interactions. Kajita's results supported the idea that atmospheric neutrinos created by cosmic rays change from one type to another in flight. This still didn't quite solve the solar neutrino problem, as they weren't looking at neutrinos coming from the Sun. Finally, on 18 June 2001, Art McDonald and the SNO collaboration announced that they had used their beautiful gold-hued detector to demonstrate that solar neutrinos oscillate, solving the mystery of missing solar neutrinos that Ray Davis had observed almost fifty years earlier.

After the Nobel Prize ceremony in Stockholm in 2015, McDonald visited the many institutes that had made the win possible. One of those was Oxford, where he celebrated among his many colleagues in the wood-panelled dining room of Mansfield College. Although I am not a neutrino physicist, I was lucky enough to attend. Between the main course and the dessert McDonald stood up to speak. 'No one encounters neutrinos in their daily life,' he said. 'Maybe once in your lifetime a neutrino will change one of your atoms and you won't even know it.' We now know that neutrinos are abundant – they are the most common particle that we know of in the Universe. Tens of billions of them go through your thumbnail every second, but they are very, very hard to detect. SNO is an extreme example of the approach particle physicists have been forced to make in order to understand particles as elusive as neutrinos.

Thanks to McDonald and Kajita's experiments, we now know that neutrinos can change type over time and distance. This is a very strange idea. Perhaps the best analogy I've come across to describe it is from Emily Conover at the University of Chicago,[13] who likens a neutrino to Cinderella riding in her carriage to the ball. She starts out in something that definitely looks like a carriage, but as she approaches the palace, the probability that her coach will turn into a pumpkin gets higher. Thinking in terms of quantum mechanics, we can say the carriage is simultaneously a pumpkin and a carriage and it depends where along its trajectory you observe it as to which it is. If Cinderella were to travel on an electron neutrino, there's a chance that by the time she gets to the ball (or a detector) she'll find herself on a muon or tau neutrino.

This oscillation requires – mathematically speaking – that neutrinos have a small mass, yet we still don't know which neutrino is the heaviest, nor exactly what each of the masses are. Other particles do not oscillate; this seems to be a property specific to neutrinos. All we know is that if you added all three masses together it would still be a million times lighter than an electron. We don't know why they are so light.

Neutrinos do not feel the strong or electromagnetic force; they only feel the weak force and gravity. From the perspective of a neutrino, matter barely exists at all; it is just a few electrons spinning around in space. This makes them very hard to detect, but it also makes them a key tool for investigating the weak interaction, while avoiding electromagnetic and strong force interference. In time this insight led to the creation of beams of neutrinos driven by particle accelerators (proton beams create pions, which then decay to muons and neutrinos), and the Nobel Prize in 1988 for Leon Lederman, Jack Steinberger and Melvin Schwartz, who first established that electron and muon neutrinos are distinct (the third type, the tau neutrino, was finally detected in a dedicated experiment in 2000 at Fermilab).

Today, we also know there are other unusual features of neutrinos that seem to set them apart from all other particles. For instance, most particles can be 'left-handed' or 'right-handed', but not neutrinos. All neutrinos are left-handed and all anti-neutrinos are right-handed. Handedness for particles means the direction of the particles' spin and how it relates to the direction the particle is travelling. If you curl your hands into fists, notice how even if you point your thumbs in the same direction (the direction of travel) the fingers of your left and right hands curl in opposite directions; that's akin to the handedness of a particle.

We haven't figured out why neutrinos don't come in both left- and right-handed varieties. What we do know is that there are many sources of neutrinos out there in the Universe. In 1987, neutrino bursts from a supernova were detected by multiple experiments, giving rise to a new field of neutrino astronomy. In a star, photons of light are constantly interacting, being absorbed and re-emitted by atoms. It can take 100,000 years for photons to travel from the core of a star to the surface. In contrast, neutrinos travel out into space unhindered, letting us see into the heart of the Sun and into supernovae in a way that other particles can't. Beyond our galaxy, extremely high-energy particles are created out in space and it is highly likely that neutrinos will one day be the messengers that teach us how those cosmic particle accelerators work. Perhaps it will give us a mechanism to copy in our laboratories here on Earth.

Neutrinos are also created much closer to home. In fact, beta decay is also going on in the interior of the Earth, producing antineutrinos.[14] The Borexino detector, an experiment designed to look for these *geoneutrinos* (in addition to solar neutrinos), is located in a lab in the depths of a mountain in Gran Sasso, Italy. The collaboration of 100 physicists from Italy, the United States, Germany, Russia and Poland are trying to figure out how much of the Earth's heat is caused by radiogenic heat, which is generated inside the Earth by radioactive decays of mostly potassium-40, thorium-232 and uranium-238. This is incredibly important to geologists since heat drives almost all the dynamic processes on Earth, from volcanoes to earthquakes, and has spurred an entirely new field called neutrino geophysics.

*

Beyond interesting new areas of science and fascinating questions in particle physics, when I set out to write about Project Poltergeist and its successors I knew I'd have to admit at this point that we currently have no direct use for neutrinos in our daily lives. Despite that, they are so important to the overall story of particle physics that excluding them would be an unforgivable oversight.

The neutrino is a classic example of curiosity-driven research that doesn't appear to have any practical applications. In comparison to a zippy electron that interacts with matter via the electromagnetic force, or a neutron that interacts with atomic nuclei via the strong force, the chargeless and almost massless neutrino is like a barely perceptible puff of a particle that interacts with almost nothing. Yet if we look back through the experiments we've seen so far, we know that it's not always obvious what the specific use of a discovery will be.

Many of the discoveries we've seen so far were premature compared to the technologies of their day: synchrotron light didn't seem useful at first, nor did the electron. The photoelectric effect wasn't fully utilised in everyday technology for decades. Particle accelerators weren't invented to produce medical isotopes or to treat cancer. No one was eagerly awaiting these discoveries except the physicists who made them, and even then the discoveries weren't always intentional. While it's likely that neutrinos will never be as directly useful as electrons, the knowledge we have gleaned from them is important and – incredibly – there are a few possible applications in the pipeline.

In the Boulby mine in the north of England a UK–US collaboration is currently building a new experiment called WATCHMAN (WATer CHerenkov Monitor of ANtineutrinos).[15] This project will use a neutrino detector to monitor nuclear fission reactors remotely by detecting the neutrino flux they create. The project could provide a unique contribution to global security by creating a reliable way of checking whether reactors are compliant with non-proliferation treaties. Because neutrinos are so hard to stop, there is simply no way of hiding an operating nuclear reactor from a detector like this.

Neutrinos may also indirectly help us transition from fossil fuels and nuclear fission reactors as a source of power to *fusion* reactors: our best current moonshot to have abundant, safe, low-carbon electricity in the future. Fusion reactors re-create nuclear reactions similar to those that power the Sun, without any potential to 'go critical', but getting one working requires that we are absolutely confident in our knowledge of nuclear physics. This knowledge has come in part from the solar neutrino experiments by Ray Davis, Super-Kamiokande and SNO, which have confirmed that our model of how neutrinos are formed in the Sun is correct.

Further in the future, there may be direct applications of neutrinos and the knowledge we have about them. Because of their ability to cover vast cosmic distances at almost the speed of light without hindrance, neutrinos could even one day become a kind of cosmic messaging system. If there are any advanced civilisations out there living on one of the thousands of exoplanets that we have discovered, neutrinos might well be the way they communicate with each other. It sounds more like sci-fi than science, but in 2012 a neutrino experiment called MINERvA (Main Injector Neutrino ExpeRiment to study v-A interactions) at Fermilab tried it out. They encoded a beam of neutrinos with a binary message using a proton accelerator, sent it through half a mile of rock to a detector, and successfully decoded it again.[16] This could also be useful on Earth, for submarines trying to communicate through water for instance, where radio waves get distorted by obstacles. With neutrinos they could communicate not just through water but also straight through the centre of the Earth in a direct line.

It's fair to say that neutrinos are not quite ready to use yet, and perhaps they never will be. We cannot predict the future, but what we can say about neutrinos is that the outcome of our quest to understand them has contributed to our lives in indirect, but profound, ways. We've already seen that SNO was located in a deep underground laboratory in Canada, which has now been expanded and renamed SNOLAB. When they say deep underground they really mean it. At 2100m underground, the laboratory is located twenty times deeper than the Large Hadron Collider, which features later in this book. The air pressure increases by 20 per cent as you take the six-minute journey down in the lift. Nigel Smith, the Executive Director of SNOLAB until 2021, describes the journey as feeling a little like descending in an aeroplane, while surrounded by rock.

The underground lab is not just host to particle physicists. Its creation opened up possibilities in many other areas of science. Being so deep in the Earth it is a unique environment because the laboratory has an incredibly low level of background radiation from cosmic rays. The existence of a stable, clean underground facility with such low radiation levels has enabled a broad research programme looking at the impact of low radiation levels on cells and organisms. No land-dwelling animals have ever lived – or for that matter evolved – without exposure to background radiation from cosmic rays, so these experiments are helping biologists understand what the impact is when you remove this radiation. This is important because it may answer the question of whether radiation is always bad for cells and organisms, whether it always causes damage, or if there is some threshold level of radiation which is harmless or possibly even beneficial to life. It could tell us more about whether evolution is influenced by the random mutations caused by radiation. So far, the results seem to indicate that life actually needs a low level of radiation.[17] If further experiments validate this, it has enormous implications not just for humans and our interactions with radiation, but also for our understanding of the existence of life elsewhere in the cosmos. Without deep underground labs, we simply couldn't do this research.

SNOLAB also happens to be one of the best places on (or in?) Earth to run experiments on quantum computers. There is emerging evidence that the decoherence time, that is the time for which a quantum 'bit' can store information before it loses it, may be limited by natural background radiation on the surface of the Earth. In the future, it may be necessary to run quantum computers underground. For now, at least, these laboratories provide a rare space for this development work.

The neutrino has been called a ghost, a messenger, a spaceship, a wisp of nothing. It started life as an apology to save a basic law of physics and over time it led to enormous payoffs in astronomy, cosmology, geology and our most fundamental understanding of matter.

Neutrinos are now part of the Standard Model of particle physics, but some of their properties – left-handedness, having a mass, changing type – have shown us that there must be physics beyond the Standard Model, which of course prompts countless questions. Why do neutrinos have mass? Are neutrinos their own antiparticle? Do the oscillations of neutrinos and antineutrinos act the same way and if not, could that explain why we see more matter than antimatter in the Universe? The neutrino, tiny as it is, turns out to be a billion times more abundant in the Universe than the matter that makes up stars, galaxies and us. It has driven experimenters and theorists alike to ever greater heights, or technically depths, to unravel its secrets. Ironically, in saving one basic law of physics, the neutrino is now one of the richest sources of knowledge gaps in physics. It affirms that there is so much about particles and forces in the Universe that we are yet to discover.

10

Linear Accelerators: The Discovery of Quarks

Along the south coast of Britain, a series of giant concrete dishes look out to sea, the largest a 200-foot curved wall. From a distance, they look like satellite or radio equipment, but they pre-date these technologies. Built between 1915 and 1930, these carefully shaped structures are sound mirrors, which were installed to provide an early warning system for enemy aircraft approaching the shore. The idea was ingenious, using large parabolic dishes to reflect sound waves to a focal point where an operator would listen for the noise of a plane's propeller. However, they were rather ineffective – which was just as well, because it wasn't long before a new technique made them obsolete.

By the late 1920s radio transmitters and receivers were starting to become mainstream, and in 1935 British physicist Robert Watson-Watt invented a system that could bounce short-wave radio signals[1] from distant moving objects like ships or aircraft and detect the reflected waves with an antenna to pinpoint an object's location. He called the system 'Radio Detection and Ranging', or *radar*. By 1939, when the Second World War broke out, a string of radar stations had been set up along the south and east coasts of Britain.

Radar promised to be vastly superior to sound mirrors, but to reach its full potential the system needed three key improvements. First, it needed to operate at an even shorter wavelength to be able to detect small objects like German U-boats. These submarines, which were attacking and sinking ships on a regular basis, could in principle be detected by high-frequency radar if they surfaced. Second, the system also needed much more powerful radio transmitters than those available at the time to reach further afield. And third, a radar system was needed that could be mounted on fighter planes, so it would have to be much smaller and lighter than existing systems. This drive to make radar work in a time of war led to enormous advances in technology, from telecommunications to cancer treatment. At the same time, these gains in radar technology would be perfected by physicists in order to make one of the most challenging discoveries ever made: quarks.

Over on the Californian coast, Stanford physics graduate Russell Varian and his younger brother, the pilot Sigurd Varian, were living in a socialist-theosophist community called Halcyon, working on their own ideas for radar technology. They tried to build a lab in the community to work on it, but in isolation it was not very successful. In 1937 the brothers decided it would help to work more closely with Russell's ex-roommate from graduate school, Bill Hansen. Hansen was an expert in radio-wave-emitting technology at Stanford University. They struck a deal with the university, who would give them no salary, but offered a $100 budget and would take a 50 per cent stake in the profits of whatever was invented from the venture.

Hansen was brought up in California and had an interest in mechanical and electrical toys at a very young age. A star student, especially in mathematics, he graduated from high school at fourteeen and enrolled at Stanford two years later, first studying engineering and then experimental physics. In graduate school Hansen worked on atomic physics, which was how he met fellow graduate student Russell Varian. Russell was often underestimated owing to his dyslexia. By this time, Hansen's interest wasn't purely in generating radio waves; instead, he wanted to create a particle accelerator for electrons.

Hansen had the idea that a metallic cavity with the right dimensions could be designed so that electromagnetic waves could resonate inside it. Then, he might be able to send a beam of electrons through and use the electromagnetic waves oscillating inside to accelerate the beam. He named his device a *rhumbatron* because of the way waves bounced around. Yet he had a similar problem to that of the radar pioneers: to make it work, he needed a source of radio-frequency power with a wavelength shorter than any existing source.

Within twelve months, Hansen and the Varian brothers had invented a device called a *klystron*. Inside the can-sized cylindrical device a low-power radio signal was applied to an electron beam that travelled through a series of cavities, as Hansen had imagined. The device didn't accelerate the electrons; instead, the combination of the cavity and passing electrons worked together to create a resonance and emit electromagnetic waves. The result was that the small input signal was amplified by the energy of the electron beam, producing high-power *microwaves* in the GHz frequency range. Counter-intuitively the name *micro*wave doesn't mean the wavelength is tiny – in fact, the wavelength is around 10cm, about 200,000 times longer than the visible light our eyes can perceive. The name was adopted because the waves produced were shorter than familiar radio waves. This short wavelength meant that the klystron itself was small and light, weighing only a few kilograms.

The klystron was not yet powerful enough to be used for radar, but it represented a huge step forward – the first device to operate in the microwave range and run efficiently and stably.[2] At least it was the first device they knew of, as they were unaware of a simultaneous invention in the UK.

On 12 September 1940, a top-secret delegation of six men, including John Cockcroft, arrived in Washington with what has been called by US historians 'the most valuable cargo ever brought to our shores'.[3] They carried a tin trunk containing a small copper device, with a series of documents outlining a number of other British inventions. The United States was at this point still neutral territory, and the plan[4] was for Britain to simply hand over these secrets in exchange for resources for development and production.

The copper device inside the trunk had been made by physicists John Randall and Harry Boot at the University of Birmingham in 1939. Their invention,[5] the *cavity magnetron*, is a cylindrical copper block with a large central hole, surrounded by other smaller holes arranged around the central hole like the petals of a flower. Electrons circulate inside the central hole of the device under the influence of a magnet and as they stream past the 'petals' or cavities they create a resonance that produces electromagnetic waves. The smaller the device, the higher frequency waves it creates: this one worked at 3 GHz, a very similar frequency to the klystron.

Both the magnetron and the klystron could produce high-frequency pulses with a much shorter wavelength than existing radar systems, which would enable radar to detect smaller objects and use smaller antennas. Both were compact and lightweight. What set the magnetron apart was that it could produce pulses at an unprecedented power level, and could be used to locate planes miles away. The British recognised the promise of the cavity magnetron, so kept the device a secret, but they lacked the manufacturing capacity to develop the technology on a large scale. With German bombing intensifying, the British government had decided to share the top-secret technology with the United States and ask for help.

The US contingent were at first reluctant to engage, but ultimately shared their own prototype radar developments which, they admitted, were reaching a dead end. They needed more transmitter power. When John Cockcroft and his colleagues revealed the cavity magnetron, it immediately provided the solution: it had output powers a thousand times higher than the klystron. As a result, the US government funded physicists at MIT to secretly create the Rad Lab,[6] which put together a lot of the theory and components needed to make high-frequency radar work – all using magnetron technology. At the time the only people with experience in high-frequency technology were accelerator scientists, so they were hired to do the work, pushing magnetrons to higher and higher output power. Bill Hansen visited regularly to teach the MIT physicists. At its peak, the Rad Lab employed 4,000 people and designed half of all the radar systems used during the war.

Companies began manufacturing magnetrons on a large scale and MIT chose the local electronics company Raytheon to help them in the development. Soon the big players like General Electric and Westinghouse were also producing magnetrons, along with smaller companies like Litton Industries, a vacuum tube company located in an industrial backwater outside San Francisco that had helped the Varian brothers build their first klystrons.

By 1945 one of these companies, Raytheon, was producing seventeen magnetrons per day for the Department of Defense when one of their engineers, Percy Spencer, noticed a chocolate bar had melted in his pocket while he was standing in front of a magnetron. He decided to try using the magnetron to cook food, first popcorn – a resounding success – then other foods, which he found heated rapidly when contained in a metal box. Raytheon filed a patent for the first microwave oven and the first commercial version, the 'Radarange', measured 8 feet tall and cost $5,000. Over time smaller, cheaper microwave ovens powered by magnetrons became the everyday household appliances we all know today. It was quite an unexpected spin-off from radar, but it wasn't the only one.

An article in the *Saturday Evening Post*[7] from 8 February 1942 exclaimed that 'The klystron beam is even more amazing than the inventors' dreams.' The article raved about telephone engineers using waves from klystrons to transmit 600,000 conversations simultaneously across the country, and television engineers doing the same with images. The military applications were not just to detect enemy aircraft or ships: 'Shot downward from an airliner, the klystron beam tells the pilot how far above the ground he is flying. Shot ahead, it warns him of hidden mountains in time to change his flight course.'

The klystron was licensed by the Sperry Gyroscope company for commercial and military applications including radar, and Russell and Sigurd Varian moved to Long Island temporarily to work on these classified projects. By 1948 the Varian brothers realised the commercial potential in TV broadcasting and telecommunications, so they left Sperry Gyroscope and moved back to California to start a company, Varian Associates,[8] that would manufacture klystrons for these rapidly developing markets.

The British Military were a major user of magnetrons for radar, and in 1953 decided to create a report ranking the quality of different manufacturers of magnetrons in Europe and the US. To the surprise of GE, Raytheon and Westinghouse, it was Litton Industries that came out on top. How, the big companies wondered, did this small firm outdo them? Litton was able to start making magnetrons for radar systems because they had the know-how from making vacuum tubes, but so did the other companies. What gave them the edge? It turned out there was another connection which gave them an advantage over the big companies: it was their connection with Bill Hansen, the klystron and his drive to build particle accelerators.

The Stanford group could not have built the first klystrons without collaboration with Litton Industries. Litton Industries had provided components to the Stanford group and discussed their manufacturing processes. It was from this experience that the company knew, for example, the importance of high vacuum in creating stable, high-power devices. They knew to create quality-control processes to ensure their devices could maintain high vacuum and that all the components remained clean during manufacturing. It was this trade secret that led to their success when they started building magnetrons.

With Litton and Varian leading the way, other high-tech companies started to grow in the Stanford Industrial park. Varian and their new local competitors caught the attention of people looking to work in highly skilled technical areas. Within a decade of starting up, Varian Associates occupied several large buildings and employed more than 1,300 people, with annual sales of $20,000,000.⁹ Thousands flocked to the area to work for the growing microwave and vacuum tube companies, or tried their luck starting their own businesses selling specialised materials, high-precision machining or other services. What started as a backwater is now the best-known technology hub in the world: Silicon Valley.

The growth of Silicon Valley in the history of tech is a complex story, but these companies created the industrial infrastructure in the area that made it possible. It was this concentration of high-tech skills that prepared the fertile ground from which the semiconductor industry grew in the late 1950s and 1960s.¹⁰ Just down the road at Stanford University, it would also enable one of the biggest physics discoveries of the century.

Like many physicists, Hansen's work had been derailed by the war, and his dream of building a particle accelerator for physics research had been set aside. After the war, designs of high-power magnetrons and klystrons were declassified and suddenly accelerator scientists around the world had the technology, industrialised and low cost, to be able to push particle accelerators to the next level. Hansen returned to his initial inspiration: the idea of building an electron accelerator, realising that magnetrons and klystrons – *radiofrequency (RF) power sources* – could feed power into a new type of accelerator. This was the full realisation of Wideröe's idea from way back in the 1920s: the *linear accelerator*.

Rather than applying high voltages as in Cockcroft and Walton's day, Hansen planned to put the particles through radio-frequency cavities for them to gain energy. He designed the system as a series of precisely machined copper cavities with a hole for the beam to travel through. These were the accelerating cavities. They would be powered by a klystron – chosen in part because it was his own co-invention – to generate electromagnetic waves. Inside the accelerating cavities these waves would oscillate in a mode in which the electric field gave a forward push to make particles go faster. He knew that if he could re-design the klystron to output sufficiently high RF power, the forward push and the energy a particle would gain as it passed through the accelerating cavities would be substantial. Linear electron accelerators now had the potential to be compact and efficient, thanks to the new RF power sources.

Hansen gathered a team at Stanford including Ed Ginzton and Marvin Chodorow, and by 1947 they had built their first 6MeV accelerator. His report to his funding body was just four words: 'We have accelerated electrons.' The linear accelerator or LINAC was much smaller and lighter than existing accelerators. Not long before, a team led by Luis Alvarez at Berkeley had built a lower-frequency proton accelerator and proudly taken a photo of their team, where around thirty people were lined up sitting side by side atop their (comparatively enormous) machine. When Hansen found out about this photo, he grabbed three of his graduate students and stood together with them, squished chest to back, holding their new high-frequency electron accelerator aloft with one hand. When completed it was less than two metres long: small, light, efficient, the way of the future. Hansen and others' research involved a two-way flow of innovation: the physicists invented new devices – the magnetron and klystron – which found large-scale real-world applications in radar. The industrialisation of these devices then helped the physicists in realising their own experimental ambitions.

Hansen dreamt of a much larger machine: a billion-volt electron accelerator which could be used for exploring the forces in the nucleus. This was the same time that the Cosmotron and Bevatron were being planned, and the drive to build large accelerators was reaching fever pitch. Hansen enlisted about thirty graduate students and thirty-five technicians to work towards the large machine. They built a series of prototypes, beginning with their original Mark I (the 6MeV), then the Mark II, reaching 33MeV in 1949. But, sadly, Hansen would never see the project completed because he was becoming increasingly unwell with a chronic lung illness. He passed away in 1949, just before the Mark II started operating. It was a shock to everyone, not least to his team. As Ginzton says, 'It was not obvious how the billion-volt machine could be completed without him.'[11]

All this innovation had happened before the theoretical developments of the 1950s gave physicists a deeper understanding of the interactions between particles and fundamental forces. In Chapter 8 we saw how large laboratories were formed to build enormous proton synchrotrons to create and study pions and strange particles. Throughout that period the new LINAC technology was being developed for electrons, which at first seemed to have little to do with understanding the strong force and the new particles being discovered elsewhere. In time, all that would change.

Once Murray Gell-Mann had wrangled the long list of particles into order with the Eightfold Way, it was clear that the strange particles were much more like protons and neutrons than they were like electrons or photons. To really understand strange particles it was imperative to understand the strong nuclear force. One way to do that was with large proton synchrotrons, but the trouble with this approach is that protons themselves interact via the strong force, making it almost impossible to isolate the strong force interactions of strange particles from the strong force interactions of the protons.

This was the key point of discussion among twenty or so Stanford physicists and engineers who were summoned to the house of German-American physicist W. K. 'Pief' Panofsky in the Los Altos hills on 10 April 1956. When they arrived, they were told they were all volunteers on a new unnamed project which had no funding. The prospect of an unsanctioned experiment piqued their curiosity, so they all stayed anyway. The idea of exploring the properties of the strong force in protons and neutrons using electrons emerged precisely because electrons interact via the electromagnetic force, but not via the strong force: they could use the electrons as a probe to better understand the strong force.

It helped that electrons were already well understood. In the 1950s Richard Feynman, among others, created the theoretical framework of *quantum electrodynamics* or QED, a way of calculating particle interactions based on a set of rules that made the calculations tractable. The method worked for photons, electrons and muons, their respective antiparticles and even for neutrinos. It didn't apply, however, to the strange particles or to protons and neutrons. The physicists figured that if they made an electron accelerator and bombarded materials rich in protons and neutrons, they could separate out the data from interactions they *could* calculate (using QED) from those they couldn't. That way, they might perhaps be able to isolate the strong interactions they were interested in. They calculated the energy that they would need, and the number came out twenty times higher than Hansen's 1GeV dream.[12] There was only one technology which could produce the beam they wanted and they were already working on it: the LINAC.

In a LINAC, the beam doesn't bend, so the electrons don't lose energy through synchrotron radiation (see Chapter 7). Creating sufficient data required as many electrons as possible, and the LINAC made such high-beam intensities possible because there was no need to wait for one batch of particles to be accelerated before the next could start its journey. The machine could use a continuous stream of particles accelerated in a straight line. It would need powerful RF sources – klystrons – but with a long enough accelerator, it might just work. Thankfully, the technology had continued to mature since Hansen's first 6MeV version. The team had achieved 400MeV in 1953, and by the time the 20GeV target was proposed at the Los Altos meeting, the Mark III accelerator was nearing the original 1GeV goal.

This ambitious new project needed a name, of course, and given the monstrous size of the accelerator – it would have to be around two miles long – they adopted the name 'Project M'. Technically, the 'M' was never defined, but the physicists mostly recall it standing for 'monster', befitting the biggest project ever attempted at Stanford. At a series of weekly meetings over the next year they threshed out ideas for the 20GeV linear accelerator to be located on the Stanford campus in Menlo Park. They summarised it in a 100-page document and submitted a request of $114 million to three different federal agencies.

Ed Ginzton, Hansen's long-term colleague who had co-founded Varian and taken over its leadership when Hansen passed away, led the design. Over a period of five years the team overcame a series of tortuous political hurdles until in 1961 the money was finally awarded. The Stanford Linear Accelerator Centre, known as SLAC, could finally get underway. Stanford University would retain leadership of the project, but it would be open to scientists from anywhere. The university donated the land and the Department of Energy picked up the tab for the accelerator. Now it was all coming together: a product of the right people, the right technology and the right location all coalescing around a common goal.

From the time they published their design in 1957 to the time the beam was switched on in 1966, further theoretical developments had taken place that helped create a strong driving force behind the SLAC experimental programme. In 1964 the Eightfold Way was upgraded into the more sophisticated quark model proposed independently by Gell-Mann and Zweig. Protons, neutrons, pions, kaons and the other heavy particles weren't fundamental particles after all, but consisted of three types of quark: up, down and strange, each of which had a specific spin and electric charge.[13] There was one extremely concerning upshot of the theory: the quarks were supposed to have non-integer, or fractional, electric charge.

Nothing other than whole units of electric charge had ever been seen in nature. How could these new particles have +2/3 or -1/3 electric charge? Even Gell-Mann wasn't sure if quarks were real entities, or just a neat mathematical trick that happened to work. If these strange non-integer quarks were the building blocks of atoms, and if in turn there really were quarks inside the protons and neutrons in the nucleus, it ought to be possible to create them and measure their properties. The search for quarks became the next great experimental challenge.

Experimentalists working at CERN quickly realised that particles with a charge of 1/3 and 2/3 should leave distinctive tracks in a bubble chamber, tracks that might have even been missed in earlier experiments. Two groups scoured through 100,000 bubble chamber photographs from previous experiments but found no evidence of fractionally charged particles. They then tried to find quarks using their proton synchrotron and bubble chamber but came away with nothing. The quarks either had a mass higher than they could create or they didn't exist. Otherwise, something else was going on.

The labs with large proton accelerators couldn't seem to liberate quarks directly by splitting the proton or neutron apart; instead they had to think of a different way to determine whether they existed. But how to do that? As it happened, the new facility at SLAC provided just the right conditions to make it work.

The 20GeV accelerator came to life in 1966, a process involving several thousand people from Stanford and elsewhere, and the search for quarks was the number one priority. An MIT–SLAC collaboration formed including Henry Kendall, Richard Taylor and Jerome Friedman, among others. The SLAC side was led by Kendall and Taylor. Kendall was an outdoorsy physicist originally from Boston, and Taylor – known for his quick wit and humour – hailed from Alberta, Canada. Friedman, the MIT contingent, was originally from Chicago, the artistically talented son of Jewish Russian immigrants. Friedman would commute to California to meet with Kendall and Taylor.

Their idea for the experiment was reminiscent of one we've already seen, when Geiger and Marsden bounced alpha particles off gold foil to discover that the atom has a nucleus. To find out if protons and neutrons have a sub-structure, the quark hunters of the late 1960s decided to use almost exactly the same method. The 20GeV electrons had enough energy to penetrate deep inside the protons and neutrons. If there were any quarks inside, the electrons would scatter from the impact and their angles and energies could be used to reconstruct what they had interacted with.[14]

If you drive along Interstate Highway 280 today, halfway between San Francisco and San Jose, you drive right over the 2-mile-long accelerator. When constructed, the tunnel that houses it was the longest building in the United States.[15] Inside is the klystron hall, full of the high-power radio-frequency devices invented by Hansen and the Varian brothers. The power they generate is transferred a few metres underground to the precision-machined copper cavities that make up the electron linear accelerator.[16] Inside, electrons ride the waves until they reach 20GeV, travelling at 99.9999999 per cent of the speed of light.

When all was ready with the quark hunters in the late 1960s, at the end of the accelerator the electron beams were bent and directed down three beam lines and into two experimental halls where they would hit – or more accurately scatter from – a target made of proton-rich liquid hydrogen. The scattered electrons would then pass through a device called a magnetic spectrometer, which measured the energy of the electrons by bending them in a magnetic field. The spectrometer was the largest scientific instrument of its time, 50 metres long and weighing 3,000 tons. It was also movable, mounted on rails so it could pivot about the target and take measurements at different angles.

In 1967 Kendall, Taylor and Friedman started doing experiments using the large spectrometer along with two smaller ones. What did they expect to see? Despite their ambitions to find quarks, most physicists at the time thought that they weren't real objects and that the proton and neutron had a kind of soft internal structure. The expectation was that fewer electrons would be scattered as the angle of the spectrometer increased. Any deviation from this could indicate there were quarks – or something else – inside. The experiment collected data to create a probability distribution, and the team pored over the results to try to interpret them.

There was a difference of roughly a factor of a thousand between their expectation and the experimental result.[17] It wasn't quite clear at first that this was evidence of quarks, but it did seem to be evidence of some kind of structure within the proton. Theorists including Richard Feynman and James Bjorken invented the name *partons* to describe these entities. It was a moment reminiscent of the earlier gold foil experiment, only this time the result penetrated even deeper inside the heart of matter: protons were not fundamental particles, and the evidence seemed to show that the partons inside – presumably a type of particle – were point-like. What does it mean for a particle to be 'point-like'? Just as for the electron, it means the particle is so small that its size can't be measured. As Jerome Friedman recalled later, 'this was a very strange point of view. It was so different from what was thought at the time that we were reluctant to discuss it publicly.'[18]

Over the next few years Friedman, Kendall and Taylor continued gathering data at different spectrometer angles, and they ran a second round of experiments using a liquid deuterium target to gather comparative data for the neutron.[19] With enough evidence, they could start to be confident in their results: the partons were indeed quarks, point-like constituents that formed the structure of protons and neutrons. We can now say that the proton is made of three quarks, two ups and one down, and the neutron of one up and two downs. The final piece of the puzzle was to confirm the idea that quarks have fractional electric charges. They were able to do so by comparing the scattering of electrons against similar data from CERN using (electrically neutral) neutrinos, which gave the physicists information about the electric charges involved in the interaction. Quarks did indeed have fractional charges.

Further analysis of the data revealed information about protons and neutrons that was even more subtle than the fact that they had quarks inside. Each proton or neutron has roughly equal parts of quarks and neutral gluons – massless particles now understood as the carriers of the strong force that 'glue' the quarks together, in much the same way as the photon carries the electromagnetic force. The three main quarks in the proton and neutron are called the *valence quarks*. Surrounding this is a 'sea' of quark-antiquark pairs, which also showed up in the data from scattering at lower energies. The proton and neutron can only be fully understood in terms of both mass and interactions by including the *sea quarks* – up, down and strange quark-antiquark pairs – as well as the valence quarks.

In the 1970s physicists began to understand the unusual properties of the strong force that binds quarks together. It is relatively weak at short distances, but extremely strong at large distances, a little like an elastic band holding quarks together. When of quarks are near each other they can move with relative freedom, but try to pull them apart and the strong force fights against you, a property called *confinement*. This traps quarks inside the proton and neutron to such an extent that if you try to pull them apart the energy you put in simply creates a new quark-antiquark pair. The bizarre upshot of this is that quarks cannot be observed alone in nature. This is why Kendall, Taylor and Friedman were successful where others were not: they found a way to observe quarks in their confined state inside protons and neutrons.

The strong force is also responsible for holding the neutrons and protons together inside the atomic nucleus, in a subtle way. At this larger distance, it is often called the *residual strong force*. The specifics of exactly how quarks interact was eventually detailed in a theory called *quantum chromodynamics*, or QCD, which we can use to help understand how the atomic nucleus is held together.

QCD holds that quarks carry a type of charge (analogous to electric charge) called *colour charge*, which has three types, labelled red, green and blue, although they bear no resemblance to colour in the normal sense. The anti-quarks have the 'colours' anti-red, anti-green and anti-blue and when quarks are combined into particles the total must be 'colourless'. Blue, red and green in combination is colourless, so if the quarks inside the proton are blue, red and green this particle is therefore 'allowed'. The pion consists of a quark and an anti-quark of either the up and down type, in a blue and anti-blue, red and anti-red or green and anti-green combination.

Protons and neutrons inside the nucleus are both overall colourless, but the quarks inside them leave a small residual effect of the strong force which, somewhat miraculously, has the effect of holding them together. Admittedly this seems like a minor detail, but it is not trivial: without the residual strong force the nuclei of atoms would not be stable and matter as we know it would not exist.

This all took a little while to establish, but what was abundantly clear after the experiments by Friedman, Kendall and Taylor was that quarks were indeed real.[20] The days of protons and neutrons being the fundamental building blocks of atoms were over.

The quark discovery was enabled by the linear accelerator, which itself required klystrons and magnetrons, which in turn had been created to provide high-power radar technology. Hansen and the Varian brothers could never have predicted the eventual outcome of their research. These interconnections between fundamental and applied science, industry and discovery are usually separate stories told by scientists and entrepreneurs. We hear the tale of discovery from the physicists and the tale of innovation and commercial success from the entrepreneurs, but somehow forget about the symbiosis between them. This coalescence can have unpredictable results, and this story doesn't end with quarks.

When we last encountered the Varian brothers, they had started their company in what would become Silicon Valley. They were soon selling electron LINACs for applications outside of physics and these machines brought transformational change to medicine, security and industry. Today the name Varian is almost synonymous with linear accelerator technology, and the product you are most likely to come across – which one in eight of us are likely to need in our lifetime – is a radiotherapy LINAC.

In 1954 a medical doctor, Henry Kaplan, heard of the accelerator developments at Stanford and moved there with the goal of creating a device to treat cancer.[21] Kaplan had lunch with Ed Ginzton and their enthusiastic collaboration led to the development of the first clinical LINAC in the United States. This 6MeV electron machine was used for the first time in 1956 in Stanford to successfully treat a two-year-old boy with an eye tumour. He walked away tumour-free, with his vision intact. Kaplan pushed to train radiologists in the new type of therapy, and the demand for accelerators in hospitals started to grow.

Kaplan and Ginzton convinced Varian Associates to manufacture a clinical accelerator. They shrunk the 6MeV machine down even further until it was compact enough to be rotated 360 degrees around the patient, allowing doctors to treat patients from any angle. From this point on, X-ray radiotherapy became the treatment method of choice, with the LINAC its method of delivery – by the time protons and heavier particles came to be clinically useful, this form of radiotherapy was already the gold standard (see Chapter 8).[22]

Today roughly half of all cancer cases are treated with radiotherapy when it is available (the rest are treated with surgery and chemotherapy). The use of electrons and X-rays is far more common than of protons or ions, partly because the tech is much smaller and cheaper. A modern clinical linear accelerator is housed in a hospital basement in a radiation-shielded room with metre-thick concrete walls. To a patient, the system looks almost the same as the proton therapy centre I described in Chapter 9, except that this time all the equipment fits inside the treatment room. In the centre of the room is the bed that the patient lies on and above them is a metre-long particle accelerator which boosts electrons up to about 25MeV and then directs them onto a metal target. As the electrons slow down in the metal they emit X-rays, just like in the cathode ray tubes we met in Chapter 1. Radiotherapy treatments rely on taking these X-rays and shaping them using a sophisticated *collimation system*, which absorbs X-rays to create a shadow pattern according to a treatment plan. Once shaped appropriately, the X-rays are directed towards the patient.

All the power supplies, vacuum systems and electronics are enclosed behind a panel at the back of the machine. Opening it up reveals the klystron and the waveguides that feed RF power to the accelerating cavity structure at the heart of the device. The so-called *gantry arm* containing the accelerator itself also has some lead shielding and a series of magnets that direct the beam downwards onto the metal target, where X-rays are created. The whole accelerator is in a plastic enclosure with imaging and control panels around it. At the press of a button, the entire accelerator system can rotate 360 degrees around the patient's bed.

Varian is one of two main players who dominate the market in medical accelerators today. The other is Elekta, formed by Lars Leksell in 1972 in Sweden based on 'Gamma knife' precision radiosurgery equipment. While Varian's machines mostly use the klystrons of their own invention, Elekta's technology mostly uses magnetrons. Both companies have many active collaborations with university groups and continuously innovate to improve their machines to achieve the best possible clinical outcomes.

There are more than 12,000 of these radiotherapy LINACs in use around the world. These machines are a reminder that experimental technology is used not just for particle physics, but to save the lives of millions of people. In fact, 12,000 LINACs is not nearly enough. With current cancer incidence rates, one machine is needed for every 200,000 people and while high-income countries meet this target, there is currently a shortfall of around 5,000 machines in nations classed by the World Bank as Low and Middle-Income Countries (LMICs). In Sub-Saharan Africa, there are thirty-five countries that currently have no radiotherapy provision at all.

Incidence of cancer is growing worldwide as a result of people living longer, and the growth is fastest in LMICs. It's estimated that by 2035 there will be around 35 million people diagnosed with cancer each year, and 65–70 per cent of all cancer cases will be in LMICs. Enormous global efforts to eradicate other diseases are working, raising life expectancy around the world, but the chance of being diagnosed with cancer increases as people age. Medical facilities are sufficiently advanced in many LMICs to make a cancer diagnosis, and increasing access to education means that people know the signs of cancer well enough that they will visit the doctor.

By 2035 there will be a need for an additional 12,600 machines, along with tens of thousands of oncologists, radiologists, medical physicists and other medical professionals. Terrific work is being done by the International Atomic Energy Agency (IAEA) to address this, but the growing need for facilities is outstripping the rate at which new radiotherapy installations are being built and commissioned.

In 2016, a meeting was convened at CERN to discuss these machines, bringing together international accelerator and global health experts with doctors from Nigeria, Botswana, Ghana, Tanzania, Zimbabwe and other sub-Saharan African nations. The technology experts spent three days listening and asking questions, seeking to understand what was going wrong and what needed to change. I was one of those experts, and once my eyes had been opened to such a global challenge, they could not be closed again.

Even if a hospital can afford a machine, the annual maintenance contract costs roughly the same as it would to pay the salaries of twenty-five full-time engineers. The spare parts can take a long time to arrive and even then can get stuck in customs for months. And each day the accelerator is broken, around fifty patients won't receive treatment. These are the most common particle accelerators in the world, but what we learned is that they are designed for high-income countries with stable power supplies, an army of highly trained engineers and strong health systems.

The participants at that meeting joined together to launch a new collaboration, STELLA – Smart Technologies to Extend Lives with Linear Accelerators. There are many aspects of this challenge to be addressed, including education, global development, health systems and technology. Using the collaboration models that underpin Big Science we aim to solve this challenge, and our first phase – to design a more suitable LINAC for these conditions – is well underway.[23]

Outside of medicine, there are many more uses for LINACs. There are thousands of small accelerators used in security scanning systems at ports and borders, allowing customs officers to take images inside trucks and cargo containers to find contraband. The high-energy X-rays produced by LINACs can get through the large objects that standard X-ray tubes can't.

Electron accelerators are used to sterilise medical products, some at-risk postal mail, and even to remove potential pathogens from certain foods including herbs. The number of applications is only growing. In South Korea you can now find small linear accelerators cleaning noxious fumes from power plants and treating effluent wastewater from factories without using harsh chemicals. It may seem counter-intuitive, but particle accelerators may well be one of the most environmentally friendly tools we have – they are even used to produce cheaper solar panels.[24] The market for these kinds of accelerators is currently around $5 billion a year and growing.

Magnetrons, klystrons and linear accelerators continue to be developed today, both in industry and in university labs, and usually in collaboration between them. These technologies are getting smaller, cheaper, more reliable and more energy efficient. Accelerator technologies for particle physics are now commonly co-developed with their applications in medicine and industry, in part because the industrialisation process can help reduce the costs of large projects, just as it did in the search for quarks.

Today, conferences focusing on new types of radiotherapy which could reduce cancer-treatment times from minutes to seconds and from twenty-five treatment sessions to just one or two are filled with accelerator physicists.[25] The physicists who work alongside medical colleagues to invent these next-generation technologies are the same ones designing particle physics experiments. They love having the ability to create immediate real-world impact, while never having to pause in their search for answers to big questions about the Universe.

But all of this would come much later. Back at the end of the 1960s, a new era of discovery emerged. As humans took their first steps on the Moon, they also made groundbreaking forays into the smallest constituents of matter. After the discovery of quarks, physicists around the world continued to revolutionise particle physics. Between 1974 and 1977, experiments at SLAC using an electron-positron collider ring called SPEAR provided evidence that there was a tau lepton – a heavier version of the electron and muon – indicating that there might be a *third* generation of matter. If that were true, then there could also be more quarks. There seemed to be no end to the mysteries that the subatomic world was producing.

11

The Tevatron: A Third Generation of Matter

We first met Robert Rathbun ('Bob') Wilson at Berkeley in the mid-1940s, when he proposed the idea of proton therapy. By the late 1960s he was no longer Ernest Lawrence's protégé but a leader in his own right. Wilson was a new kind of physicist, a kind of jack-of-all-trades who was simultaneously a visionary, an engineer, a fundraiser and an entrepreneur. He also happened to be a gifted poet, sculptor and orator, and in time he would learn to weave the creative and scientific together to launch a world-leading laboratory. But first he had to get the funding approved to build it.

In April 1969 Wilson stood before the US Congress asking for 250 million dollars to build the most ambitious accelerator project the United States had ever attempted. The boom days of free-flowing finance for physics were over and Wilson had to compete with many demands on government money, from NASA's space missions to the enormous costs of ships, aircraft and weapons for defence. Before Wilson could even start, Senator John Pastore pointed out that the proposed machine was experimental. They didn't even know exactly what they would find with it. How could he defend such a costly and risky proposition?

The machine was designed, Wilson said, to find the answers to age-old questions about the simplicity of nature. Was it possible, he asked, that we could find a description for the messiness of all of life and the Universe, based on just a few elementary particles? From this vision he laid out the state of play. They knew about the forces of gravity, the electromagnetic force and the nuclear force which binds protons and neutrons together. The discovery of quarks was underway at SLAC and as we have seen in previous chapters, there were also hints at a fourth force, the weak nuclear force. In beta decay, during which a neutron turns into a proton, it seemed that quarks experienced both the strong and the weak nuclear force. This new machine, he argued, would allow experiments in an energy range where physicists might be able to finally confirm these forces and piece together a fuller understanding of how the Universe worked. In intellectual terms the promise of this endeavour was enormous.

Senator Pastore nodded and said that he understood the purpose of the machine was fundamental high-energy physics research, an educational and academic process. Wilson added: 'And a cultural one, but with the firm expectation that technological developments will come … Because we are doing extremely difficult technical things, and because we are working in a strange kind of research, we know from past experience that new techniques inevitably develop, techniques which have paid, more than paid, for the cost of the basic research that was not pointed to such developments.'[1]

The senator wanted to help Wilson by making it seem as if the machine was indispensable to the nation. He asked Wilson whether the machine would have anything to do with the security of the country, but Wilson simply replied 'No'. After the Manhattan Project his days of contributing to defence through physics were very much over; this project was purely one driven by curiosity about the Universe. The senator pressed him on it, 'Nothing at all?'

Wilson paused, then he looked at the senator and said, 'It has only to do with the respect with which we regard one another, the dignity of man, our love of culture. It has to do with: Are we good painters, good sculptors, great poets? I mean all the things we really venerate in this country and are patriotic about … It has nothing to do directly with defending our country except to make it worth defending.'[2]

The budget was approved. In October that year Wilson personally dug a shovel into the soil at the site an hour outside Chicago, marking the official groundbreaking ceremony of the National Accelerator Laboratory (NAL) complex, later known as Fermilab.

*

Fermilab is truly unlike any other physics laboratory. True to Wilson's interests, instead of drab brick and prefab buildings, the site is full of sculpture and points of architectural detail. Driving in through a small village of wood-panelled houses and onto the Fermilab site, visitors are met not with high-tech equipment but with a herd of bison, a nod to its prairie heritage. Approaching the main building, visitors drive between long reflective pools. At their far end is the cathedral-like[3] Wilson Hall, a 250-foot-tall concrete structure softened by architectural curves. The top floor viewing deck looks down on kilometres of tunnels and technology, accelerators and experiments spreading out like grassy crop circles across the vast site.

Wilson's vision was to create a laboratory that was exciting, functional and beautiful. He believed the aesthetics of the site would be important to its success. A full-time artist, Angela Gonzales, was hired as a core member of the team to design everything from the lab logo and meeting posters to tables in the cafeteria. The same aesthetic went into the scientific equipment. Wilson felt that scientific instruments should be just as beautiful as the ideas from theoretical physics. As a sculptor he insisted that accelerators, experiments and all the other aspects of a large laboratory should have graceful lines, well-balanced volumes and intrinsic aesthetic appeal.[4]

At first Wilson sketched out the facility with a broad brush, almost as if he was outlining forms on a canvas. He had to be scientifically ambitious in order to attract the best people to the project, but he also needed to be frugal to stick within budget. He decided that his original goal, the one that had been funded, was not ambitious enough. Instead of the original 200GeV beam energy, he wanted to aim for 500GeV with a machine – the 'Main Ring' – which had a radius of 1km, a size chosen simply because it was easy to remember. If that wasn't enough of a challenge, he also accelerated the construction schedule. Instead of the original seven years, he wanted to build the project in just five.

The best minds in the business found out about his outlandish idea and started joining the project. His vision brought in physicists, engineers and problem solvers with enormous creative energy and drive. The new Main Ring wasn't even the only accelerator that needed to be constructed. Wilson knew it needed a whole pre-accelerator chain: the protons would start in a Cockcroft-Walton accelerator, then they would be sent through a LINAC and passed to a ring called the 'booster'. Only after that would the beam of protons be passed into the Main Ring.

Accelerator physicist Helen Edwards and her husband Don joined the team in 1970, just as things were kicking off. Hailing from Detroit, Michigan, Edwards became interested in science and mathematics while attending a girls' school in Washington DC and, despite struggling with dyslexia, mastered subjects using sheer concentration. She earned a Bachelor's degree in physics from Cornell, the only woman among a dozen men. She intended to go directly into a PhD programme but women at that time were required to complete a Master's degree first. Still, she persisted, and completed her research on particle decays, gaining hands-on experience operating the Cornell electron accelerator for her experiments. This was where Wilson met her, and it was clear to all that her ability to concentrate on the essentials made her a formid-able solver of scientific and technical problems. Wilson put Edwards in charge of commissioning the booster synchrotron.

Edwards and her team quickly got the booster accelerator working, delivering 8GeV protons on demand to the Main Ring. The teams man-aging the Cockcroft-Walton and LINAC met their goal to get those up and running. With construction happening at breakneck speed, Edwards joined the team working on the partly completed Main Ring.

The pace of work was feverish, and the conditions grim: water leaks meant the Main Ring tunnel would sometimes fill with mud, which they simply had to wade through in order to continue with the magnet installation. Wilson took risks, saying 'something that works right away is over-designed and consequently will have taken too long to build and will have cost too much'.[5] It would be cheaper, he argued, to fix the parts that failed.

Problems had to be solved quickly, and Edwards' team would later tell stories of how she could produce detailed calculations at the drop of a hat to solve complex issues as they arose. Her team was no less inventive. After welding together the accelerator they found there were small bits of metal debris in the beam pipe which risked derailing the protons, causing radiation or damaging the machine. In desperation, one engineer trained a ferret named Felicia to pull a string through the beam pipe, after which he attached a cleaning swab and pulled the string back through to remove the debris.[6] It worked, but there was worse to come.

In 1971 things came to a head after the team started powering up the 1,014 magnets in the tunnel and found that no less than 350 of them had failed. This 'magnet crisis' cost them at least six months and \$2 million to fix, and even to this day it's not entirely clear what went wrong – but thin epoxy insulation and problems with condensation seem to have been mostly to blame. Despite all the challenges, in March 1972, less than four years after the site had been a cornfield, a proton beam was finally circulating round the 6.28km circumference of the Main Ring, and before long they broke the world record for proton beam energy.

Next Edwards oversaw the difficult process of getting the high energy proton beam *out* of the machine. She needed to keep at least 98 per cent of the beam or risk creating radiation and destroying components. The solution she adopted[7] involved finely tuning the machine so that the beam would veer very close to the edge of the pipe in three locations, where her team installed electrostatic septa – metal strips held at a very high voltage – which provided just enough force to coax the wobbling beam out of the machine.

By 1974 she had things working and all three experimental areas could receive the beam at the same time. In the years following, the energy of the Main Ring was pushed up from 200GeV, to 400GeV in 1975, and later to 500GeV. Each of the exit-points from the ring were split three ways again to provide nine beams in total from the one accelerator. Now that the machine was ready for use, attention could turn to the experiments sitting downstream.

The main experimental areas focused on neutrinos, mesons and protons respectively. The experiments were mostly designed and implemented by university scientists, not the lab's own staff. This was partly because of Wilson's frugality. To save costs he had decided the experimenters would be in charge of their own areas, and they would simply be provided with a 'pit' – literally a dirt-floor pit dug into the ground for radiation shielding – complete with corrugated iron walls and a roof. Clearly Wilson's aesthetic plan for the site didn't extend to user comfort. University scientists felt unfairly treated because they had to work in the gritty experimental areas at Fermilab as opposed to the comparatively plush labs at SLAC or CERN.[8]

Despite the frontier feel of the experimental pits, Wilson knew that if he created the highest energy beams in the world physicists would come knocking, and sure enough, they did. By 1976 the lab had proposals from more than 120 research centres, including international collaborators from Canada, Europe and Asia. Over half of the 500 proposed experiments were approved, and by 1978 many of these had been completed. One of the early experimental leaders was the charismatic Columbia University physicist, Leon Lederman.

Since Fermilab's inception, Lederman had been an advocate and supporter of the lab. He matched Wilson's idea of a perfect experimenter – ambitious but willing to adapt. Where Wilson was a cowboy, Lederman was a city slicker. Born in New York to Ukranian-Jewish immigrant parents, he'd chosen physics at college when a friend convinced him of its splendours over a long night drinking beer. He had developed a knack for choosing important physics problems, which led him to co-discover the muon neutrino in 1962. This set him on the path for his experiments at Fermilab.

Lederman and others knew of the two existing generations of matter, which could be grouped into the *leptons*, consisting of the electron and its heavier cousin the muon, together with the electron neutrino and muon neutrino, which formed two matching sets. Add to this the *quarks*: if the up and down quarks were a set, it seemed reasonable that the strange quark might also have a partner, named the *charm* quark, composing a second quark generation. This was proposed by theorists in 1970, originally motivated by aesthetic considerations. It also helped them resolve some technical problems in their equations.

By the time the Fermilab Main Ring was running, Lederman had already missed out on discovering the charm. It was found in 1974 almost simultaneously at Brookhaven and SLAC in the form of the 'J/Ψ' particle, pronounced 'jay-sigh' after the Greek letter Ψ 'psi'.[9] But nature wasn't done with surprises. As we saw at the end of the last chapter, researchers at SLAC found an even heavier version of the electron and muon in 1975, called the *tau*. Lederman had a new motivation: if a third generation of leptons existed, why not a third and heavier generation of quarks too?

Lederman applied for space to conduct a new experiment – named after its proposal number Experiment 288 (E288) – in which he hoped to use electronic detectors to look for pairs of muons, which were the signature expected from a short-lived, heavy particle. His goal was to find particles containing even heavier quarks than the up, down, charm and strange found so far. When his proposal was accepted and the experiment prepared, the 500GeV proton beam from the Main Ring was directed at their pit and the team gathered data on each pair of muons they found. To analyse the results, they added up the energy of each pair of muons and charted them as points on a histogram. A peak, or bump, in that histogram would be evidence of a new particle.

By 1976, a bump at a mass of around 6GeV appeared. Although the number of events was small, the likelihood of these events being a statistical fluke was just 2 per cent. They went ahead and published a paper announcing a brand new particle, which they gave the name upsilon, which they interpreted as meaning 'the lofty one'.[10] Then the unthinkable happened. As they collected more data, the peak showing the upsilon particle disappeared, swallowed up by the background noise of random events. There was no 6GeV particle after all.

It was a brutal lesson in statistics, and one of the reasons why today the commonly accepted standard to announce the discovery of a new particle is what physicists call '5-sigma' evidence. This means that the chance of producing the result by fluke is less than a one in 3.5 million chance.[11] There is almost no other area in life or science where we apply this incredible standard of evidence. For example, if you were diagnosed with a disease and the doctor said she was 95 per cent sure the clinical trial data for the treatment she was offering wasn't a fluke, you would take the medicine, right? But particle physicists would not accept this as evidence. Working on such long timescales, on such large projects, this is just one way particle physicists ensure they don't kid themselves about what is real and what is not.

Lederman took the failure with good humour, even after his colleagues re-named the non-existent particle the *OopsLeon* in his honour. The E288 team returned to their experiment in the spring of 1977 and began collecting new data. When a peak emerged at around 9.5GeV after just seven days one of the physicists, John Yoh, exclaimed, 'What the hell is happening?' But as tradition dictated he put a bottle of champagne in the fridge labelled '9.5', just in case.

This time they didn't rush to make an announcement. They were determined to make absolutely sure that this new particle, produced once for every 100 billion protons colliding with the target, was not a fluke. They buckled down to collect more data. At 11 p.m. on 20 May the cabling of a current-measuring device on a magnet failed. The cable heated up, melted, then started a fire in the nearby cable tray. Before long acrid smoke filled the hall. The team panicked.

The fire brigade arrived and quickly put the fire out, but soon the team were even more alarmed: the water they used combined with chlorine gas released in the air by the fire to form acid that began to corrode the electronic components of the experiment. If they couldn't stop the corrosion, they would never collect enough evidence to announce the new particle. Desperate to save the experiment, Lederman called in a Dutch fire salvage expert who arrived in seventy-two hours bearing buckets of a secret cleaning solution. Everyone on the E288 team, the staff in the Proton Department, husbands and wives, friends and secretaries joined the physicists in a production line to dip, scrub and clean the 900 circuit boards under the careful watch of the expert.

With the experiment salvaged, five days later they started running and taking data again. This time the 9.5GeV bump held up. The new particle had a mass around ten times larger than the mass of the proton. They double- and triple-checked their results, but this time it was definitely a better-than-5-sigma result.

On 15 June 1977 they called a seminar in the Fermilab Auditorium and announced that they had really done it. E288 had discovered a totally new particle at 9.5GeV, the heaviest particle ever discovered and the first to be found at Fermilab. They reused the name 'upsilon' but this time the name – and the particle – stuck. The 9.5 bottle of champagne was downed in celebration and Fermilab was firmly on the map as a lab for experimental discovery.

Fitting the new experimental evidence in with developments in theory didn't take long. The upsilon turned out to be a combination of a 'b quark' and an 'anti-b quark', where b stands for 'bottom' or 'beauty' depending on who you ask. The new heavy b quark had been predicted as early as 1973 by Japanese theorists Makoto Kobayashi and Toshihide Maskawa and the names 'top' and 'bottom' had been coined in 1975 by the Israeli physicist Haim Harari. Despite the increasing complexity of particle physics, the upsilon confirmed that an underlying simplicity was emerging, a theory which indicated that nature had a pleasing symmetry to it. There were six leptons (electron, muon, tau and their neutrinos) and six quarks (up, down, strange, charm, bottom, top).

With hindsight the upsilon was, as Lederman said, 'one of the most expected surprises in particle physics'.[12] Now that they knew the bottom quark existed, it followed that its heavy partner the top (t) quark must exist as well. While they didn't know quite how heavy it was, as the theory didn't tell them that, the next step at Fermilab was inevitable. They would go after the sixth – and final – quark.

Fermilab was living up to Wilson's vision as a national and international facility, but he wasn't done yet. He'd always had his sights set far beyond this first stage. By the time they discovered the upsilon, Fermilab no longer had the largest accelerator in the world. That title had gone to CERN, who built a 7km-long ring called the Super Proton Synchrotron that reached an energy of 450GeV. Wilson and Edwards had proudly outmanoeuvred them by getting to 500GeV with the marginally smaller Main Ring, but now Wilson revealed the plan he had been harbouring all along.

From the start Wilson had been looking beyond the Main Ring to bigger opportunities and he had two ideas in mind. First, he realised that if a second accelerator made up of powerful magnets was added to the complex, they would be able to double the energy of the beam. They could reuse the same tunnel to create beams at 1,000GeV or 1TeV, reaching the 'tera-scale' and potentially opening up a whole new realm of discovery. Second, he wanted to build a machine that could smash particles directly into each other instead of into a fixed target: a particle *collider* rather than just an *accelerator*.

The new ring, dubbed the 'Energy Doubler' but later known as the 'Tevatron', would lie directly underneath the existing Main Ring, where Wilson had ensured enough space was left for it. The plan was to accelerate protons first in the existing Main Ring, then transfer the beam into the new Tevatron ring to reach 1 TeV. Holding these higher-energy particles on track required a completely new type of magnet technology, one which could create a magnetic field more than double that of the Main Ring. Conventional electromagnets made of iron and copper wouldn't cut it, so Wilson had planned to use *superconducting* magnets, so-called because they're made from materials that can sustain an enormous electrical current without generating heat.

Superconducting materials lose all electrical resistance below a certain temperature, usually at around -270 degrees Celsius, an effect discovered for the first time back in 1911. Fifty years later the first superconducting materials[13] that could be shaped into wires were discovered. The wires could, in theory, drive a strong magnetic field. The problem was no one had ever made an accelerator magnet in this way. As ever, Wilson was one step ahead and had kick-started a programme to figure out how to build superconducting magnets in 1972, an astonishing move five years before Lederman and the E288 team had found the upsilon.

The second aspect of Wilson's bold vision involved colliding two beams together, and this was no less challenging. Colliding particles head-on is an almost impossible task, since each individual particle is so small that its chance of hitting another one head-on is minuscule. But Wilson pushed for this because it would open up incredible opportunities for discovery. In previous accelerators where high-energy beams were smashed into fixed targets, energy conservation dictated that most of the energy in the beam went into smashing particles out of the target and carrying them away. Only a limited portion of that energy goes into making new particles. In particle physics we call this the *centre-of-mass energy*, and in the case of a 1 TeV beam striking a target there is just 43.3 GeV of useful particle-creating energy available. Any particles with a mass higher than 43.3 GeV simply cannot be created.

It was this quantity that Wilson wanted to improve. In a head-on collision, all the incoming energy goes into the centre-of-mass energy, so two colliding 1 TeV beams have a centre-of-mass energy of 2 TeV, a feat which would require an accelerator hundreds of kilometres in circumference to achieve in traditional target-based experiments. No matter how difficult, the advantage of the collider was clear.

The original Main Ring had certainly been challenging to build, but this new idea seemed almost madness. There were risks in every single aspect and not a single part could be purchased 'off the shelf'. The new ring needed 774 niobium-titanium superconducting dipole magnets cooled in liquid helium to bend the beam in a circle and an additional 216 quadrupole magnets to focus it. They would also need spares when inevitable breakages happened. No companies knew how to build the magnets and neither did Fermilab. Alvin Tollestrup, one of the lead designers, presented the idea to their European colleagues at CERN, and he later recalled 'a big room with these guys sitting up there laughing … they thought we were nuts.'[14] The CERN physicists knew that no one in the world was capable of building the magnets the Tevatron needed and certainly not in the quantity that was required. How do you make something that has never been made before at a scale that seems almost ridiculous when not a single company is capable of producing it?

The first step was to find the raw materials for the magnets. In 1974 only a few specialist companies sold niobium-titanium superconducting material for high-tech users, and most orders were for a few grams or at most a few kilograms. The Fermilab team asked how much it would cost to buy by the tonne. Later that year, they placed an order to purchase a staggering amount of niobium-titanium, equivalent to 95 per cent of the material that had ever been produced.

With the raw material on its way, the next stage was to figure out how to make it into cables. Many had tried and failed, but at the Rutherford Appleton Laboratory in the UK a team had come up with a way of pulling the precious niobium-titanium material into very fine filaments and embedding thousands of these filaments in a copper matrix to form a strand of wire, which finally formed a cable. It sounds easy once you know how, but Fermilab had to skill up.

Once they figured it out, the lab decided that they would outsource production to manufacturers, handing over the raw materials and the recipe for how to achieve long, perfect lengths of wire. Rather than patenting the wire-making procedures, the Fermilab team decided to make them freely available, opening up competition for manufacturers to supply the finished cable for their enormous project. Once the cable was made, it could be formed into coils and powered up to create a magnet.

All this care and precision was necessary to make sure that the superconducting wires didn't *quench*, which is where tiny heat effects can cause a magnet to lose its superconducting state and to suddenly heat up. A quench isn't just a minor setback. It releases an enormous amount of energy, and if you don't manage it correctly, the magnet and its power supply can simply explode. Superconducting magnets are an extremely delicate business.

Today we have the hindsight of decades of research, but in the 1970s no one had a clear recipe for how to build this kind of magnet and there was very little theoretical understanding of the complexities involved. Wilson, a practical and skilled craftsman himself, realised the challenge ahead of them and decided to create a 'supermagnet factory', placing Tollestrup in charge of development. The magnets were so sensitive to changes that Tollestrup decided only one variable should be changed on each attempt. There was no other way forward but through trial and error.

Between 1975 and 1978 they built around 100 foot-long (30cm) magnets, each with a slightly different design, trying at each step to figure out what worked. If a short prototype showed promise, they would build a longer version, until eventually they reached a full-length magnet of 22ft (6.7m). Experience proved that any slight changes to the construction method could spell disaster. Just because a magnet worked well when it was a shorter length didn't guarantee it would work when it was longer.[15]

Fermilab's R&D method was certainly considered unusual at the time. They created the manufacturing know-how to scale up the production from individual working magnets to almost a thousand units, and they did it all in-house to ensure they could control fine details and achieve the necessary quality and consistency. Ultimately, they had to be sure that every magnet was exactly alike, avoiding any imperfections or magnetic differences, either of which would have a disastrous effect on the proton beam. Only after this intensive, multi-year effort could the magnets be strung together into a ring and turned into a particle accelerator.

This magnet development work was all going on while the upsilon was being discovered, scrubbed from the record books and then discovered again. Wilson had divided the accelerator staff effort and asked some of them to focus solely on the Main Ring, while others worked on the Tevatron. Among those growing frustrated was Helen Edwards. Along with a few others, she had concerns about the Tevatron, so together they formed an unofficial group called the Underground Parameters Committee to work on the design aspects that troubled them. Wilson found out and decided to support their work, from a distance.

But technical problems were not the only concern. Fermilab was running short of money and the government had yet to approve funds for the Tevatron. In 1978, in the midst of funding shortages and with the Tevatron vision not yet realised, Wilson grew increasingly frustrated with his role as director. Eventually, he decided to walk away, handing over the role of leading Fermilab to Leon Lederman, who had to make a decision – and quickly – on whether they would proceed with the new ring or cut their losses and turn the Main Ring itself into a collider. Competition from CERN was relentless. They were already adapting the Super Proton Synchrotron into a collider with 270GeV in each beam, giving 540GeV centre-of-mass energy with which to search for heavy particles.

Lederman organised a review meeting in November 1978, known as the 'shoot-out' debate. Proponents for and against building the Tevatron presented their arguments, and experts from other labs were brought in as judges. They grew convinced that they couldn't compete with CERN by building a collider with the Main Ring. They also gradually became assured that despite all the risks the Tevatron was feasible. During the course of this debate a second advantage of superconducting magnets became obvious. At a time when oil prices were skyrocketing and electricity shortages abounded, Fermilab's electricity bill had reached about $10 million a year, a huge chunk of the lab's operating cost. The superconducting magnets could be energised and then left running, and this would save the lab around $5 million a year on their energy bills.

By the end of two solid days of debate Lederman had made the decision: they would proceed with the Tevatron. The newly formed Department of Energy (DoE) agreed to the plan in stages. The Fermilab team had to show strings of magnets operating reliably first in a test hall and then in the main tunnel before the project would be allowed to proceed.

The job of leading the design and construction of the Tevatron accelerator was an enormous challenge, and Lederman put Helen Edwards and Rich Orr jointly in charge of it. Orr was a physicist who hailed from Iowa, known for his calm demeanour. He'd helped build the meson lab and like Edwards had become well known for his skills in bringing people together and encouraging them towards success. Together they were a formidable duo who knew how to make the team their first priority, which would turn out to be critical in a project the size of the Tevatron.

The magnet test went off without a hitch. It was so successful they decided to push the magnets hard, running the current up to 4,000 Amps, trying to cause quenches. The quench protection systems all kicked in beautifully, letting out boiling helium and protecting the magnets. They tried to cause electrical arcs but, as Rich Orr recounted later, 'we couldn't break it'. The magnets were ready to go. Production ramped up, the magnet factory went into full output mode and workers could be found in the tunnel around the clock, running pipes, making connections, doing electrical work and installing magnet after magnet.

By mid-June 1983 they had the beam running in the Tevatron ring. Two weeks later, on 3 July, they reached a beam energy of 512GeV, a new world record. They had taken the lead from their European rivals and newspapers heralded their success. But for Edwards and Orr an even greater challenge was ahead. They were about to attempt to turn the machine into a collider, smashing a proton beam into an antiproton beam.

Ideas for colliders had been around since at least the 1950s and 1960s.[16] The first one to be built was a small electron collider called AdA (Anello Di Accumulazione) at Frascati in Italy in 1961. CERN built the first proton collider called the Intersecting Storage Rings (ISR) in 1971 and it achieved a 60GeV centre-of-mass energy. With almost forty times the energy of the ISR, the Tevatron would collide protons and antiprotons on a scale far beyond what had come before.

To get a collider working properly requires a lot of technical ingenuity. The density of a particle beam is lower than a solid or liquid target, so the beams have to cross many times and as many particles must be packed into each beam as possible. Once the protons and antiprotons were in the ring, it took about twenty seconds to get the beams up to 1 TeV, when they would be steered with magnets to cross paths in two locations. Finally, when all the pieces came together, on 30 November 1986[17] the first proton and antiproton beams slammed into each other. The accelerator physicists had succeeded in delivering the impossible: the world's largest superconducting accelerator. But where their work ended, the work of the experimental particle physicists began.

By the early 1970s the many discoveries we have seen so far were brought together mathematically to form one overarching theory called the Standard Model of particle physics. The Standard Model includes all the particles we've seen discovered, from the electron, muon and tau and the neutrinos, to the quarks and their composite particles: protons, neutrons along with pions, kaons, resonance particles and so on. However, there was one quark left to find: the top quark. It was expected to be heavy, so finding it would need collisions with as much energy as possible. This is what motivated the experimentalists who built the Tevatron.

International teams of physicists got to work setting up two major experiments, which involved building two enormous detectors at the points along the ring where the Tevatron beams collided. The first experimental team built the Colliding Detector at Fermilab (CDF) and elected Alvin Tollestrop and Roy Schwitters as co-spokespeople. The CDF collaboration came together quickly, with physicists from the University of Pisa in Italy and the University of Tsukuba in Japan joining colleagues from around ten US institutions. CDF was a huge 4,500-ton layered cylindrical detector embedded in a superconducting solenoid to bend the particles and allow their momentum to be determined. The different detector layers were sensitive to different particles, allowing the eighty-seven scientists who worked on it to measure particle energy, charge and type, and to create digital reconstructions of the debris from particle collisions. All the layers were now completely electronic, so data acquisition and computing became integral to the experiment. To build the detector each collaborating institution took responsibility for a different part of the detector and took charge of the financial and technical aspects of constructing and delivering it. Eventually it was all brought together and started gathering data in 1986.

After CDF a second detector, DZero (named for its location in the ring), was built. The DZero team had some catching up to do, but eventually the collaboration grew to a similar size to CDF, and both ended up with several hundred collaborators. Two experiments were necessary in order to ensure independent verification of any new physics. DZero was slightly more bulky than CDF. Weighing in at 5,500 tons, it was over four stories high, with layers of detectors similar to CDF. DZero started taking data in 1992.

These two incredible devices constituted a new type of particle detector that wrapped around the beamline. The detectors were so complex and so costly that they could not be dismantled at the end of an experimental run, as with previous accelerator experiments. Instead, they had to become multi-purpose detectors because they had to stay in place. The scale of experiments that the physicists planned to run in this new collider were unprecedented and could last longer than it takes to get a PhD or work on a tenure-track faculty job. Even the leader of experiments would just take the reins for a while before handing them over to another colleague. This was no longer just Big Science; it was megascience. Fermilab went from being a national user laboratory to a truly international one, with researchers from many parts of the world joining the programme.

Two experiments were necessary in order to ensure independent verification of any new physics. By the end of the 1993 Tevatron run, the two experiments cautiously started talking about evidence for a sixth quark, the top quark, but they needed more time and more data to reach the crucial 5-sigma level of evidence. Finally, in 1995, the two teams announced the discovery of the top quark. They had found the last matter particle of the Standard Model, by far the heaviest fundamental particle ever found. A top quark weighs more than a gold atom, despite being a point particle like an electron. It lasts for just half a trillionth of a trillionth of a second (5×10^{-25} seconds), decaying to the next heaviest quark, the bottom quark.[18] The top quark is so short-lived it doesn't have time to combine with other quarks, so unlike other quarks, which always combine together, the top quark spends its incredibly short lifetime alone.

It was a two-decade journey from the 1970s discovery of the bottom quark in the upsilon to the momentous discovery of its partner – the top quark – and the achievement made headlines around the world.

It's hard to overestimate just how difficult it is to find a particle like a top quark, as its creation is extremely rare even among the collision debris of the Tevatron. To do so, experimental particle physicists had to be experts not just in hands-on experimentation, but also in statistics and computational methods. This was quite a different skill-set from their peers even twenty years earlier. In large part this is because particle interactions are probabilistic in nature, as dictated by quantum mechanics. Not everything about the experiment could be calculated by hand, and there was no sense in building an experiment that would be incapable of finding the top quark, or other particles and processes they were looking for, so preparation was essential. So how did they do it? By using computer simulations, physicists could input all the known theoretical information and probabilities involved and then use an approach known as 'Monte Carlo' simulation to get an overview of the statistical outcomes of the experiment.

The name for the technique comes from the famous 'Monte Carlo fallacy', or gambler's fallacy, which gets to the heart of the idea that while a single event may be unpredictable, the outcome of many events can be well defined. The story goes like this.

In 1913 at the Monte Carlo Casino in Monaco, a roulette ball fell on black twenty-six times in a row. The chance of this happening is one in 66.6 million, but the chance of falling on black on each spin is always the same, 50 per cent. With each new spin, the gamblers were convinced that the next ball would surely fall on red. As the number of spins landing on black increased, from eight, nine, ten and more, they became so sure that it *must* fall on red on the next spin that they bet millions of francs. They lost it all. The only guaranteed way not to lose money betting on these kinds of statistical games is to keep increasing your bet each time you lose, so that when you win you recoup your previous losses. This is not only psychologically extremely difficult, but also is usually not allowed by the casino. They cap the size of the bet you can place, so they always win.

The predictable outcomes of these roulette wheel spins inspired mathematicians including Stanislaw Ulam and John von Neumann back in 1946, when Ulam was working at Los Alamos. His team had run into a challenge where they needed to calculate the diffusion of neutrons in a material. They knew the average distance a neutron would travel before it hit an atomic nucleus, and they knew how much energy was involved in the collision, but despite that, they couldn't calculate the answer mathematically. Ulam was in hospital recovering from surgery and trying to figure out the chance of a successful hand while playing solitaire when he had an idea. Why not run a large set of trials – like a roulette wheel spin, coin toss or solitaire hand – and play out what happens in each case? By tracing the outcomes of a large number of neutrons, where each follows a different series of collisions defined by known probabilities, the overall neutron diffusion could be determined. One of Ulam's colleagues dubbed this technique the Monte Carlo method.

Over time, as computational power grew, these methods became more and more powerful. The general idea is to avoid doing incredibly long – or even impossible – calculations by hand and instead to perform a large number of random trials. Particle physics has stayed at the cutting edge of these developments, so that by the time the Tevatron was built, physicists were already using sophisticated computer-generated Monte Carlo simulations to design detectors, model the outcomes of experiments and much more.

By doing so, experimental particle physicists can create datasets that look very similar to what they expect their experiment will create. They can develop algorithms to analyse the expected data before the experiment is even built, which allows them to check the uncertainties associated with it and to see if the experiment has a chance of producing a statistically significant result (and knowing how finicky we are about statistical significance, this is worth the effort!). If a theoretical model exists for a particle or interaction, they can even generate the 'signal' they are looking for and hide it in the background to check if the analysis algorithm finds it successfully.

This arduous preparation phase means that physicists can run their analysis algorithms as soon as they have real experimental data to see if there is a difference from the simulation. If there is, that's a prime indicator that we've found new physics. When creating an experiment to find rare interactions like the creation of a top quark, this is the best method we have of ensuring that we will find small signals among all the known physics effects. Among the uncountable billions of particle collisions that happened at the Tevatron, the physicists were thus successful in identifying the few dozen top quarks produced per year.

This high level of statistical training that physicists receive can have some unusual consequences. Over drinks at a Fermilab conference once, some US colleagues shared a story with me about the 1986 American Physical Society conference, the largest physics meeting in the United States. The conference had to find a new venue at short notice to host a whopping 4,000 physicists. Naturally, they chose the city that hosts more than 21,000 conferences a year, Las Vegas. Rather than gambling, the physicists preferred to gather around tables with a free drink and paper and pen to sketch or calculate as they talked. Without discussing it, they collectively made the only gambling move that was guaranteed to win; they didn't gamble at all. As a result, the hotel suffered its worst financial week on record. The conference was such a disaster for the hotel that at the end of the week the City of Las Vegas officially asked them never to return. From what I can tell, the story is true.

Folklore aside, the statistical literacy and expertise involved in Monte Carlo simulation techniques mean that particle physicists are extremely well placed to build models of processes and systems outside of physics, and are highly sought after to do so. Monte Carlo simulations are used for everything from weather forecasting to financial modelling, in telecommunications and engineering, in computational biology and even in law. Many of my peers from my student days have gone on to work in consulting, banking, climate-change modelling and epidemiology. I remember many of these friends who were making the transition out of physics expressing genuine surprise that the level of computational and statistical training of their new colleagues involved only basic spreadsheets.

The Tevatron was an ambitious project in many ways, but one of the most impressive side-effects was in the area of superconducting magnet technology. As early as the 1940s other physicists had realised that strong magnets could align hydrogen atoms inside the human body, and that by using a particular sequence of magnetic fields and radio waves they could measure different materials inside the body, including the locations of individual hydrogen atoms. The technique was originally called 'nuclear magnetic resonance' or NMR, later renamed 'magnetic resonance imaging' or MRI. When it was first invented, there was no way to make magnetic fields strong enough for it to be useful and commercially viable. The Tevatron changed all that.

Fermilab's ambitious project created the demand and knowledge that industry needed to produce commercial volumes of high-quality superconducting wire. Two main manufacturers had been involved: Intermagnetics General Corporation (IGC) provided 80 per cent of the wire and Magnetic Corporation of America (MCA) the remainder. Other suppliers emerged as the high-energy physics community adopted the technology more widely. At CERN they were developing a large bubble chamber using superconducting magnets, while in the field of nuclear energy large magnetic confinement devices called tokamaks used superconducting wire as well.[19] The market took off and superconducting magnets became available for widespread use.

Today, commercially available MRI scanners are used to take images of the inside of the human body, particularly of soft tissues. They are a complimentary technique to the CT scans we've already covered, but MRI is unique in that it doesn't deliver any ionising radiation to make images. You'll now find them in most major hospitals in developed countries where they are used to enable better and earlier detection of many types of cancer, in addition to taking images of the spine, heart, lungs and other organs. In the last five years, MRI scanners have even been combined with radiotherapy accelerators (see Chapter 10) to make an 'MR-Linac', allowing image-guided therapy and the ability to alter the treatment dose depending on daily changes in tumour shape, size and position.[20]

As well as hospital-based medical usage, you'll also find many MRI scanners in research labs. The technique of functional MRI is able to show where blood flows in the brain, which indicates areas of brain activity. This has facilitated a revolution in understanding the brain, the nature of consciousness and the formation of memories. It has also led to the discovery of neurotoxins that are flushed from our brains during sleep, which could lead to understanding how to help sufferers of Alzheimer's disease.

There is now a $10 billion per year and growing global market in MRI scanners.[21] The applications of MRI alone more than prove Bob Wilson's original argument when he said before Congress that there would be long-term spin-offs from the Tevatron. Although decades of research had to coalesce before this was possible, the investment now seems more than justified. Of course, the Fermilab physicists cannot lay claim to the invention of MRI techniques. But without the magnetic innovation required to build the Tevatron, superconducting technology in hospitals might never have become a reality.

Superconducting magnet technology has also been adopted in areas outside of particle accelerators and MRI. Physicists at Brookhaven had success with their 1968 patent on the concept of Maglev, a transportation technology that utilises superconducting magnetic levitation and is now used in some of the world's fastest trains. Superconducting magnets are also used in power generation and transmission, experimental fusion reactors and energy storage systems. As Robert Marsh of Teledyne Wah Chang, now the world's largest supplier of superconducting alloys, said, 'Every program in superconductivity that there is today owes itself in some measure to the fact that Fermilab built the Tevatron and it worked.'[22]

By the time the Tevatron had found the top quark, physicists in the United States had a bitter pill to swallow. For all their success, they were now forced by their own government to pass the baton of world prominence in high-energy physics to Europe.

While Wilson and Edwards had been inventing the Tevatron in the mid-1970s, theories had started to emerge pointing to new ideas with attention-grabbing names like supersymmetry, technicolour and string theory. All of them were consistent in predicting that there would be something beyond the energy reach of the Tevatron. What's more, the Standard Model, the culmination of decades of research in high-energy physics, was still missing a final piece. The Standard Model contained a prediction that they needed to confirm or deny, the Higgs boson, a force-carrying particle (with spin-0) whose mass was unknown. Leon Lederman tried to drive home the importance of this missing piece of the puzzle, coining it the 'God particle'.

It was easy to see the direction high-energy physics was heading. All of these ideas required accelerators that could collide particles at energies higher than the Tevatron. But the budget the US government was willing to shell out for the adventure was fixed or even shrinking. The United States had watched Europe join together to build CERN, where they had just started excavating an enormous 27km-long tunnel crossing between France and Switzerland to house their next machine. The next logical step was to leapfrog CERN by joining forces globally and building what they called a World Laboratory.

Leon Lederman was a strong proponent of building a new machine with global partners to help pay for it, and the idea emerged of a collider twenty times as powerful as the Tevatron as early as 1976. They sketched out an 87.1km-long ring made of superconducting magnets, to collide two 20TeV beams, called the Superconducting Super Collider (SSC). They reassured the government this project would put the US firmly back in the lead of high-energy physics. A project of this size would bring prestige and even boost the local economy, creating an estimated 13,000 jobs. Fermilab wanted the machine in Illinois, but the state of Texas won the bidding process and a site in Waxahachie, 48km south of Dallas, was settled on. The project was endorsed in 1983 and by the mid-1980s the excavating teams had started creating the enormous tunnels. Physicist Roy Schwitters from the University of Texas, Austin, who had led the CDF experiment, became project director.

Then the problems started. To tackle such a large project the Department of Energy tried to impose a military-industrial style of working which didn't sit at all well with the scientists. The scientists were accused of mismanagement and failure to control the budget and schedule. Trust began to erode. In 1987 the project was audited, and a heated debate ensued about the high cost of the project, estimated at $4.4 billion. It was the same, they argued, as NASA's contribution to the International Space Station. Yet unlike the Space Station the SSC was not living up to its vision as a World Laboratory. The nationalistic rhetoric of the United States leading high-energy physics had gained the favour of its government, but didn't sit well with their global partners, including Canada, Japan, India and Europe. When push came to shove, none of the partner countries committed to funding the project, except for a $50-million pledge from India.

By 1992, the United States was in recession and the budget had ballooned to $12 billion. The Soviet Union had collapsed, so there seemed to be little reason to prove US superiority with mega-projects. Congress wanted to end the scheme. At this point, 22.5km of tunnel had been built, $3 billion had been spent, and the buildings to house scientists and workshops had been constructed. Two thousand people were employed, hundreds of them scientists who had followed the vision of the enormous project and moved their entire families from countries like Japan, India and Russia. They believed it was too big and too far along to be cancelled. At the eleventh hour, Bill Clinton stepped in to try to save the project, arguing to Congress that they would be ending more than three decades of prominence in basic science.

In the end, nothing could be done. Congress decided to end the project, and on 1 October 1993, Clinton regretfully signed the bill that killed the dream. The Superconducting Super-Collider was nothing more than a partially completed tunnel in the ground. There were many lessons learned from the downfall of the SSC, but to me, the major one is that Big Science was no longer a nationalistic tool to dominate on the world stage. Collaborating nations expected to be treated as partners, not little siblings, in such huge endeavours. Had the SSC been a genuinely international World Laboratory as it had originally been envisioned, perhaps things would have worked out differently. The buildings were later purchased by a chemical manufacturer named Magnablend and today the underground tunnels collect rainwater. It's rumoured that some entrepreneurs now use it as a dark, humid site to grow organic mushrooms.

Despite its successor's demise, the Tevatron itself was a remarkable project which paved the way for superconducting accelerators around the world. As a result, the last quark in the Standard Model – the top quark – had been found. While all this was happening, physicists at CERN had discovered heavy force-carrying particles called W and Z bosons, which helped to cement their understanding of the weak force, and solidify the Standard Model of particle physics. Finally in 2000, the Tevatron also found the tau neutrino, completing the list of matter particles in the Standard Model. Yet there was still one undiscovered piece of the puzzle missing: a force-carrying particle called a Higgs boson.

Besides this missing piece, theorists and experimentalists were at a precipice: theorists pointed to physics *beyond* the Standard Model, while experimentalists had the tools and confidence to go after even the most difficult-to-find particles with large colliders. Could a theory of everything finally be within grasp? They just needed the right collider to find out. The Tevatron experimentalists created a programme to start looking for the Higgs boson starting in 2001, with Tevatron 'Run II', but for many others the focus shifted across the ocean to Europe. At CERN a machine which built on the superconducting legacy of the Tevatron had been constructed. The Large Hadron Collider was about to come to life.

12

The Large Hadron Collider: The Higgs Boson and Beyond

On 10 September 2008, physicists around the world were eagerly awaiting the start-up of the biggest machine on Earth, the Large Hadron Collider. The LHC is a 27km-long circular proton-proton collider built at CERN near Geneva, 100 metres underground beneath the border between France and Switzerland. The machine had started as an idea in 1984, was approved by CERN Council for construction in 1994 and after two and a half decades in the making, it was about to accelerate protons for the first time. For Welsh physicist Lyndon (Lyn) Evans – who had overseen the LHC as project director since it broke ground in 1994 – it was the culmination of his entire career building particle accelerators.

Evans comes across as modest, kind and practical.[1] This modesty extends to rarely mentioning his own career in interviews, but his nickname gives him away a little. 'Evans the atom' is the elemental force behind the LHC. Evans has worked at CERN since 1969, but his career has taken him wherever his expertise was needed, tracking higher and higher energy colliders: from the 300GeV Super Proton Synchrotron at CERN to the Tevatron at Fermilab and the Superconducting Super Collider in Texas. When that was cancelled, the LHC became the future of particle physics and bringing the project to reality was Evans's own raison d'être: 'there is nothing bigger than that a scientist could hope to achieve', he said.[2]

The magnitude of his work is not lost on Evans, who still feels overawed every time he goes into the LHC tunnel.[3] His work to build the collider involved overseeing around 2,500 CERN staff, along with another 300 scientists and engineers from Russia, China, the United States and other countries who built components for the accelerator. Evans recalls that when he met the President of China, he thought to himself, 'Not bad for a bloke from Aberdare!'[4]

Unfortunately, not everyone was as excited about the LHC startup as Evans and his colleagues. In the build-up to the big day certain news outlets were filled with headlines like 'Boffins "won't destroy Earth"' and had been busy peddling the idea that the LHC might create a black hole and destroy us all, fuelled by an ill-informed legal case trying to stop the machine from starting. This kind of conspiracy happens every time a big new accelerator starts up. 'Big Bang machine could destroy Earth' read a headline in 1999, back when an accelerator called the Relativistic Heavy Ion Collider started up in the United States. Of course, it's running fine.

Cosmic rays are bombarding the Earth from space all the time with much higher energies – thousands of times higher than the Large Hadron Collider beams – and have been doing so for Earth's entire five-billion-year existence without a problem. The difference is that the LHC creates these high-energy collisions intentionally, on demand and much more frequently than cosmic rays. The entire particle physics community had pinned their hopes of discovery on these collisions: they were looking not just for the Higgs boson – the missing piece of the Standard Model – but for anything nature might present that lay *beyond* our current understanding of physics.

Around the circumference of the LHC are the four main experiments called ATLAS (A Toroidal LHC ApparatuS), CMS (Compact Muon Solenoid), ALICE (A Large Ion Collider Experiment) and LHCb (Large Hadron Collider beauty experiment). Their goals cover almost all the big questions in particle physics, from the existence of dark matter to the reason we see more matter than antimatter around us. The construction of the LHC and its experiments were quite different journeys. Where the LHC was 80 per cent provided by CERN, with around a 20 per cent contribution from partners, the experiments – consisting of enormous particle detectors – were the opposite. The experiments were built by autonomous collections of scientists around the world who came together to form large international collaborations, with CERN providing about 20 per cent of the effort including the underground caverns and infrastructure.

ATLAS is the closest experiment to the main CERN site in Meyrin, one that visitors sometimes get to see if they are lucky enough to go on a tour. The entrance is an unassuming warehouse door; visitors shuffle through the cathedral-sized hall towards an enormous circular hole in the floor. Darkness is all that can be seen over the barrier. Above is a heavy metal crane, its rugged steel used to carry whole trucks down into the depths of the Earth. Every piece of the ATLAS experiment was lowered down through shafts like this and pieced together like an enormous, liquid-helium-cooled ship in a bottle.

Visitors pass through an industrial-looking blue metal cage to arrive at a silver elevator door. Here, everyone must put on a blue hard-hat with 'CERN' emblazoned across. Next comes the descent: 100 metres down into the Earth. Excitement builds as you step out of the lift onto a metal walkway that clangs underfoot. Around a corner is a wall which extends multiple storeys above and below. Only it isn't a wall, really. It's covered in cables and electronics and soon you realise it's a series of concentric layers of detector technology. This is the ATLAS detector. At 46 metres long and 25 metres in diameter it's often compared to the size of a cathedral, yet the numbers cannot prepare you for the reality of how huge it is. Visitors try to comprehend its many layers, from the pixel detectors giving precise particle tracks in the middle, out to the muon chambers at the edge, which catch particles that might pass through the first layers undetected.

The visit continues down a staircase, following the beam pipe, which heads off into the distance through a concrete shielding wall. Arriving in the 3.8m-diameter tunnel, visitors come to one of the 10m-long superconducting magnets, painted blue. The eye is drawn down the length of this magnet and on to the next one; beyond lie more than 1,500 of these behemoths in a 27km-long tunnel loop, the Large Hadron Collider itself. The bend of the circle is so gentle that the machine appears to fade away infinitely into the distance. Seeing it close up only makes it seem more unreal, more confounding in its complexity.

When I first saw ATLAS and the LHC, I was an undergraduate summer student at CERN. I was working on 'a control system for the heaters of the cooling system of the ATLAS inner detector'. The title says it all, really. It was far from the grand physics challenge I had envisioned, but I quickly decided that the insignificance of the project didn't matter. What mattered was that I was there, that I had the chance to participate in one of the greatest experiments ever built. At this point, the machine and its detectors were still under construction, so we were sent on tours to see all the underground parts of the experiment first-hand. This was far more exciting than my project, until I realised that within the code that I wrote – poorly – was an alarm message that could be sent up the chain of command to turn off the entire ATLAS detector. After seeing the experiment first-hand, my project suddenly seemed more important.

By 2005 there had been twenty years' worth of summer students, interns, temporary staff and others who had contributed together with the work of thousands of physicists, engineers and professionals. If a novice like me was allowed to send signals that could shut the machine down, if my mistakes or poor programming could break the entire enterprise, surely the laws of statistics said the whole thing would never even switch on.

Three years later, in 2008, I was watching the experts in the CERN control room. For start-up day, CERN did what any open, transparent organisation ought to do: they invited journalists in to watch the machine come to life. As a result, Evans ended up narrating the switching on of the largest and most complex machine ever built, with the whole world watching. A tape barricade separated the journalists from rows of computer screens arranged in circular pods, each of which controlled different aspects of the enormous experiment. Only a few specialised members of the team were allowed at the controls, responsible for bringing the world's largest accelerator into operation. Evans was with these experts and they seemed, understandably, a little nervous.

There was apprehension as the day started. Overnight some of the cryogenic systems had almost thrown a (metaphorical) spanner in the works. By morning all had settled and the start-up was given the green light. Evans oversaw the process we call 'beam threading', repeatedly passing tiny quantities of injected beam through the thousands of magnets one section at a time, correcting the orbit on each attempt so that subsequent protons remain centred in the beam pipe. At 8:56 a.m.

UK time, the cameras focused on one of the many computer screens showing blips from beam position monitors, little electrical signals from the wobbling path of the beam as it made its way through the magnets in the ring miles away from where the operators stood. Reporters were saying the proton beam had travelled more than 6km around a section of the ring. A few anxious faces turned to smiles. Two minutes later, with remarkably few tweaks, another bit of proton beam made it halfway round.

At three-quarters of the way, twenty minutes later, it reached the ATLAS detector and spontaneous bursts of applause started breaking out. In the background, cameras overheard one of the team saying 'I think I'm going to win my bet: an hour'. By 9:24 a.m., they had threaded the beam all the way around the ring once. Now the applause erupted in earnest. They had done it.

For the accelerator team, it was a triumph. Paul Collier – the British physicist who was head of the CERN accelerator department – summed up the relief and exhaustion, saying, 'I feel as if I've pushed the particles round myself.' It had all gone much more smoothly and faster than expected. I was in awe; these experts had somehow pulled it off despite the odds, creating a machine that worked beautifully, just as designed.

If the picture in your head of a proton beam is one of a laser-like state of organisation, I can assure you that's not true – in reality it is closer to the messy, complex formations we see in the early stages of galaxies. The particles in the beam are not passive participants in their relativistic joy ride nor in their eventual cataclysmic demise. Every single proton interacts with all the others and with their surroundings. Each proton in the LHC is swirling, pulling and pushing electromagnetically in its 27km-long universe, forming into one of 2,808 bunches spaced just twenty-five nanoseconds apart. Precise magnetic and electric fields create these nanoscopic particle galaxies and keep them lapping the ring 100,000 times per second for multiple days at a time, until they are eventually collided together. At top energy, if the beams veered off course or escaped, they could turn 600kg of solid copper into a pool of liquid. As you can imagine, getting all this right requires the brightest minds, the most advanced computer simulations and the best engineering that exists anywhere on Earth.

By the end of start-up day, Evans and the LHC team had beams circulating in both directions, and the experiments including CMS and ATLAS were starting to see beam splash events – not from beam collisions, but from high-energy particles bumping into a few motes of residual gas left in the beam chamber. One by one, the detectors came alive and responded by lighting up with particle tracks. The spokespeople for each experiment rushed over from their control room to the main control centre (about a twenty-minute drive) to deliver bottles of champagne wrapped in hasty printouts of the first electronic tracks from their beautiful detectors.

In the next few days, the camera crews left, the orbits were stabilised, the final focusing system designed to squeeze the beam for collisions was brought online, and all was on course. The path ahead was to accelerate the beam up to the multi-TeV range before heading towards first collisions. Then, nine days after it first started, the LHC blew up.

The experts said a 'serious incident' had occurred. As they ramped up the strength in the magnets – a normal procedure – one of the superconducting joints between two magnets had a short circuit. That is not normal. The superconducting wire fell out of its superconducting state, creating a sudden electrical resistance to around a million amps of current which released heat. A lot of heat. That evaporated six tonnes of liquid helium into gas, expanding so dramatically that the release valves that had been designed for such a scenario simply couldn't cope. The blast physically ripped almost thirty magnets, weighing 35 tonnes each, out of the floor. As images came through of the tunnel, it was carnage. Insulating material had been blown apart and debris had tainted kilometres of beam pipe. The one saving grace was that nobody was hurt, save for a few thousand egos.

It took them nine months to recover the machine, replacing the damaged magnets with spares, bulking up connections and enlarging every single release valve so that such an accident could never happen again. They dug into the details forensically to understand what had happened, sharing everything openly at conferences. Even though superconducting synchrotrons had been built before, this incident really brought home one of the difficulties of doing something so large and so difficult: the LHC is its own prototype.

The repairs were successful and the machine restarted in 2009, moving through the stages of commissioning and eventually ramping up to the full 7 TeV energy in each beam. During operation the machine has proved itself to be an elegant beast, but operating it is no less challenging than on that first day. Keeping the beams on course is an intensive task involving both electronic and human feedback systems. The operators regularly have to correct for unbelievably small effects including the movement of the Earth's crust due to the Sun and Moon, the amount of water in Lake Geneva and the timing of passing TGV fast trains, all of which affect the proton orbit. Despite all this, there have been no other major incidents in more than ten years.

In building the LHC, CERN and the international collaborations behind each of the experiments had to create epic quality-control systems to ensure everything would work reliably once it was down in that subterranean tunnel. It was truly only in writing this book that I realised the code I contributed as a student would have been passed to a professional, checked and perfected to rigorous standards before it ever had a chance of being used. The whole enterprise of the LHC is an absolute triumph of project management, of engineering and of collaboration.

The LHC has continued operating smoothly ever since, working seamlessly, 24/7, together with its chain of injectors – the series of accelerators that feed beams into the LHC. Altogether, this enormous system delivers two beams of hundreds of billions of protons to 99.999999 per cent of the speed of light, focuses them down to less than the width of a hair, and then collides them together. So what comes next? Physics, of course.

By the time the LHC started operating, the Standard Model of particle physics, the comprehensive 'theory of almost everything except gravity', was complete in its theoretical details. As we have seen, the Standard Model includes the matter particles: the 'leptons' including the electron, muon and tau and their three corresponding neutrinos, and the six quarks (up, down, strange, charm, bottom, top). The matter particles seemed to come in three generations – each generation is almost identical except for increasing mass – and the experiments to find and confirm the particles of the third generation of matter had filled in the gaps. As we saw in the last chapter, the top quark was found in 1995 and the tau neutrino in 2000 at Fermilab, the last matter particles to be discovered.

In addition to the matter particles, the Standard Model contains the bosons or 'force carriers'. We have to ignore gravity for now, as it is not included in the Standard Model, but the other three forces – the weak, strong and electromagnetic forces – are all included. The electromagnetic force is mediated by the photon. The strong force, which binds the quarks and the protons and neutrons together, is mediated by gluons. The weak nuclear force is a little bit different. Unlike the photons and the gluons, which are massless, the W and Z bosons found at CERN in the decades leading up to the LHC were actually extremely heavy.[5] The weak force also came with some other subtleties.

At high-energy scales (which we now know are above 246 GeV)[6] the electromagnetic and the weak force are actually parts of one overarching force, the electroweak force. Although these two forces seem very different on everyday energy scales, at very high energies like those shortly after the Big Bang (before quarks even formed), the two forces are mixed up with one another and cannot be separated. This was confirmed at CERN using the predecessor accelerator to the LHC, called the Large Electron Positron collider (LEP), which put the Standard Model under intensive tests like never before. This is one of the reasons physicists sometimes refer to particle colliders as time machines that re-create the conditions of the Big Bang, since they can produce interactions with energies as high as those found in the very early Universe. Experiments had also concluded that there were only three types of neutrinos and, as a consequence, that there are only the three generations of matter, as least as far as we know currently. The Standard Model seemed to be correct to an incredible level of precision. Yet still there remained a missing piece: the theoretical particle that could give the heavy W and Z bosons their mass, known as the Higgs boson.

This new particle had been predicted back in 1964 in three separate papers, one of which was written by Scottish theoretical physicist Peter Higgs. The theory postulated the existence of a field (the 'Higgs field') throughout all of space. At high energies (where the 'electroweak' force is one force) all particles are massless. At some critical energy scale, reached as the Universe cooled, the Higgs field grew and particles started interacting with it, thereby acquiring a mass. This irreversible process is known as 'spontaneous symmetry breaking' and its consequence is that different particles have different masses because they have different levels of interaction with the Higgs field.

What does it mean for the Universe to be filled with a Higgs field, and what does that do? A prize-winning explanation[7] of this is to imagine a room full of socialites at a cocktail party. If a regular person walks into the room, they can traverse the cocktail party without hindrance. But imagine if a famous person enters the room. The socialites – the Higgs field – gather around the famous person – a particle – slowing down their progress across the room. A famous person who is slowed down a lot is like a particle which is given a heavy mass by the Higgs field.

Showing that nature really obeys this Higgs mechanism required physicists to create and detect the characteristic particle that the theory predicted, the Higgs boson: an excitation of the Higgs field. This is like a rumour spreading across the cocktail party, causing socialites to clump together and transmit the excitation. In colliders, the super-high-energy particle collisions can rattle the Higgs field. Doing so causes particles to pop out of the field – these are the Higgs bosons. The only problem was, the Standard Model gave no indication what mass a Higgs boson should be. Higgs bosons were going to be fiend-ishly difficult to find.

The search for the Higgs at CERN had already started back with the LEP collider. After the other scientific goals had been met and the Higgs was the only remaining piece of the Standard Model to be found, the LEP detector collaborations turned their attention to this most elusive particle. Just before it was shut down in 2001, all four experiments had tantalising hints of a Higgs boson at a mass of roughly 114GeV, but they didn't have enough data to form any conclusions. It seemed that LEP simply didn't have enough energy to create a Higgs boson, if one was even really there. They had to let the Tevatron team at Fermilab take the lead in the hunt for the Higgs, but only for a while. CERN's long-term strategy had always been to use the tunnel created for LEP to secure the future of the labora-tory into the twenty-first century. In 1984, five years before LEP was even running, CERN had already started designing the next step, a high-energy proton-proton collider – a discovery machine that would take them far beyond the 2TeV reach of the Tevatron and up to a centre-of-mass energy of 14TeV. This was the machine that became the Large Hadron Collider.

*

It would take much more than the hardware of the accelerator and detectors to find the Higgs boson. By the LEP and LHC era – and now the projects take so long to develop we can talk of eras – particle physics was very different from its earliest days. The detectors were made of layers of specialised sub-detectors, functioning like giant layered digital cameras with millions of information channels. With more collisions than ever before, and more resolution to detect the debris from the collisions, the volume of data from the experiments was creeping ever upwards. When LEP started operating in 1989, the calibration data quickly ran into the gigabytes and experimental data into the terabytes.[8] This doesn't sound like much today, but in 1989 a standard hard drive had a capacity of just a few tens of megabytes. Where was all the data going to go? How would people access it?

This 'computing problem' became a real danger that would have to be solved. Always on the front foot, CERN networked their computers and mainframes together and started communicating by email (yes, even before the 1990s!), but what still didn't exist was a way to collaborate and access data reliably. It was at this point that Tim Berners-Lee, an Oxford physics graduate working in scientific computing at CERN, proposed bringing together new technologies in computers, networks and hypertext into a system that could help solve this challenge. He wrote a short paper outlining his idea called 'Information Management: A Proposal', on which his CERN boss scribbled 'Vague but exciting…'

What Berners-Lee invented was the World Wide Web. Yes, *the* World Wide Web. Berners-Lee came up with three key inventions you probably see every day that underpin the web: HTML, HyperText Markup Language, which is the formatting language for the web; URLs, Uniform Resource Locators, which are the unique addresses used to access each resource on the web; and HTTP, the HyperText Transfer Protocol, which is the communications protocol used to connect servers and send information. By 1990 Berners-Lee had published the first website and made the first web browser. The rest, as they say, is history.

Today there are more than 1.6 billion websites and 4.33 billion active users worldwide. That's 57 per cent of the total human population. The average user spends an astonishing six and a half hours online every day.[9] Although the internet (the physical network part) existed before the web, it is really the web we mean when we talk about 'using the internet'.

It is impossible to put a value to the web and almost unthinkable to imagine going back to an era when it didn't exist. Over time, society has simply adapted to the ubiquity of its existence, but an example might help us put it in context. In 2019 the Indian government cut internet access to Kashmir in a bid to reduce public protests. Even in this poor region, the effect was immense. Students trying to take online exams could no longer get international qualifications; e-commerce was ruined and factories selling goods were cut off as they couldn't communicate with buyers; and hospitals and pharmacies couldn't order medicines to treat patients. The cost to the economy in the nine months following the imposition of the blackout was estimated at $5.3 billion[10] out of a total GDP of around $17 billion. The ban was lifted slightly in 2020 owing to the coronavirus pandemic, but at the time of writing is still not fully restored.

Berners-Lee realised very early on that in order for the web to flourish, he would have to set it free. As he says, 'you can't propose that something be a universal space and at the same time keep control of it'. In April 1993, when there was a total of just 600 websites worldwide, CERN decided to place the World Wide Web software into the public domain without royalties or patents.

The web was a completely unexpected spinoff from physics. The needs of particle physicists and their collaborative way of solving difficult problems meant they had to share data in ways far ahead of other areas of society. As a result, it took just one creative leap in the right supportive environment to create one of the most important inventions in our modern world. Today, Berners-Lee is the director of the World Wide Web Consortium, which continues to oversee the development of the web. In 2012, when London hosted the Olympic Games, the opening ceremony featured Berners-Lee sitting at a small desk live-tweeting the words 'THIS IS FOR EVERYONE' which lit up the stadium seats like a giant LED screen. Ironically, the US TV commentators had no idea who he was, and encouraged their viewers to Google him using the very same technology he had invented.

With the LHC, CERN's data challenges increased exponentially. Although computing power and connectivity have increased, so too has the amount of data produced by the experiments, requiring constant innovation. The annual data output of the LHC detectors was forecast to be around 90 petabytes each year, equivalent to 56 million CDs – a stack that would reach halfway to the Moon. Providing all that computing power, along with storing and processing data at CERN, was out of the question – the electricity costs alone would be prohibitive. CERN's computing experts knew that eventually the datasets would be simply too large to transfer or handle in their entirety, as the copper cables that make up most of the internet simply wouldn't be fast enough.

In response, they created an international collaboration to form a globally distributed network of fibre-optic, super-fast connections and enormous computing centres, connecting scientists around the globe. Called the Worldwide LHC Computing Grid (WLCG) but referred to as 'the Grid', this system has more than 200,000 servers located in collaborating nations around the world. It can be used both to store and to process data and has successfully enabled the international collaboration that is so essential to the success of CERN.

With all these computing and engineering challenges in hand, the Large Hadron Collider came back online in 2009. It was soon colliding beams and collecting data from each of the experiments, and the availability of the Grid meant that analysis ramped up quickly. Each day the baton was passed from time zone to time zone and analysis continued twenty-four hours a day somewhere in the world. Colleagues in Australia could tap into and analyse the same LHC data as physicists in Europe, the United States and elsewhere. But they weren't the only ones on the trail of the Higgs boson.

At this point, the Tevatron was well into Run II – which had started in 2001 – and had been upgraded with the Higgs in mind. The physicists at Fermilab knew they couldn't reach the same energies as the LHC, but they were hoping they could find the Higgs first if the mass of it happened be a kind of 'Goldilocks Higgs', one which was not too heavy (>180GeV) since then they'd be unable to create it, and one which was not too light (<140GeV) since it would decay into bottom quarks and be lost in the noise. As the LHC gained energy and its collision rate improved, their Higgs-finding capabilities were gaining fast on the Tevatron.

The Tevatron teams worked furiously analysing their data. By early 2011 they were able to rule out masses up to 103GeV and between 147GeV and 180GeV with 95 per cent confidence. Just a little more data, they begged, and they might find the Higgs in between.[11] Yet budget cuts were on the horizon and the Tevatron was due to shut down in September 2011. By July the LHC experiments ruled out the range from 149GeV to 190GeV, but in September, unable to find the $35 million a year Fermilab would need to continue running, the Tevatron was shut down. In a fitting end, Helen Edwards oversaw the ceremony to put to sleep the giant machine she had worked so hard to coax into life almost three decades earlier. Now all eyes turned to the LHC.

By December the range of the Higgs had been narrowed down to 115–130GeV, focusing in on an area at 125GeV where both ATLAS and CMS saw a hint of something exciting. It was only at a 2-sigma level, and they would never forget the OopsLeon – but it had been seen independently by two experiments. The excitement among the physics community was palpable.

In July 2012, after three years of the LHC running and a fervent period of analysis, the world's particle physics community converged at a major conference called the International Conference in High Energy Physics (ICHEP), which was held in Melbourne, Australia. CERN held a press conference from their site near Geneva, streamed live – via the web of course – to the auditorium in Melbourne where most of the physicists were. I was watching the feed from my office at the Rutherford Appleton Laboratory, down the road from Oxford, joined by millions of people who watched online around the world.

The spokespeople for the two main experiments, physicists Joseph Incandela from the United States for CMS and Fabiola Gianotti[12] from Italy for ATLAS, made their presentations on behalf of thousands of scientists. I was impressed at the level of scientific detail they provided despite the presence of the media. As they each showed the reconstruction of the different Higgs decay channels, my mind was reeling with just how much work was behind every single plot and number shown.

As I watched, I was thinking of my colleagues for whom this day was the culmination of decades of work. Some had offices just down the corridor, and some were on the other side of the world from me in Melbourne. This was the work of individuals coming together in small teams of around ten to fifteen researchers, which each took ownership over their own small piece of the puzzle. These teams then joined to form larger teams or working groups with other institutes, which then coalesced into the collaboration of around 2,000 scientists per experiment. They were all working together in a self-organised management system which is the hallmark of CERN's collaboration style. Unusually, that day they had been sworn to secrecy, but we all knew what was coming.

When the physics presentations were done, it was German particle physicist Rolf-Dieter Heuer's turn to take the stage in his role as CERN's Director General. After a few preliminary words, he took a deep breath and announced 'we have a discovery'. Cheers erupted, physicists hugged and congratulated each other. They had moved countries, uprooted their families, worked countless hours at unsociable times and wondered, all along, if there was even anything there to find. Now, they had done it. They had discovered the Higgs boson. The camera zoomed in on eighty-two-year-old Peter Higgs as a tear rolled down his cheek.

When we stand back from that moment and look at all CERN has achieved, the international cooperation alone is staggering. The LHC experiments involve 110 different countries, including CERN's twenty-three member states and eight associate member states, observer nations and countries which have co-operation agreements (like Australia). It involves roughly half of the 13,000 particle physicists in the world. Even as an established scientist who regularly works in collaborations across multiple time zones, it is still difficult for me to fathom the workings of such an enormous global team. Just getting to the start, to the first beam, then to those first collisions, was a remarkable feat, let alone the success of a major discovery.

As the example of the web shows, CERN does things differently from other large organisations. CERN is funded by taxpayers' money, so almost everything it does is open source. They are promoters of the ideas of open science, open data and open hardware. Even the gift shop has to obey this rule: it cannot turn a profit. The web grew out of these principles of sharing and openness, with no idea of how it would eventually evolve. This unique aspect of how CERN works has not been missed by policymakers and international organisations.

In 2014 CERN celebrated sixty years of science for peace together with the UN. CERN is an example of how nations can work together for the global public good. Following CERN's model a number of other projects have now built similar collaborations, bringing together countries across deep political divides. The SESAME (Synchrotron Light for Experimental Science and Applications in the Middle East) centre in Jordan brings together Bahrain, Cyprus, Egypt, Iran, Israel, Jordan, Pakistan, the Palestinian Authority and Turkey. In south-east Europe, SEEIST[13] (South-East European International Institute for Sustainable Technologies) is a knowledge-economy building project which will focus on a new proton and carbon-ion therapy and research facility. CERN also helped bring about one of my own collaborations, STELLA (Smart Technologies for Extending Lives with Linear Accelerators) where together with collaborators in Sub-Saharan Africa, we aim to improve access to high quality cancer care globally by finding technological solutions to the severe shortfall of radiotherapy facilities.

These kinds of initiatives and collaborations are essential to our global future. The CERN model creates a mechanism for international co-operation with unrivalled potential to solve global challenges. Today, the UN and CERN are working together to learn how to create collaborations to make progress on the Sustainable Development Goals, many of which need science and technology solutions, including addressing climate change, healthcare and access to food and water.

There is no way that CERN would be able to have the impact it has if it were a single-country think-tank or company designed to create technology patents. The same ethos that created the web has also created a huge drive to encourage scientific research and to make its results more open to the public.

Of course, the web isn't the only technology spinoff from CERN. For those with commercial potential, there is a whole Knowledge Transfer team to develop them. Anyone can see CERN's current technology portfolio online,[14] and some current examples include collaborative software systems, radiation-hard detectors used in medicine, and compact orbital cutters to cut huge pieces of pipe in the field. The unique requirements of

CERN's large experiments have continually pushed industry to innovate in order to supply state-of-the-art components. In a survey, 75 per cent of suppliers to CERN noted they had increased their capacity to innovate through contracting with CERN. They also talk about the 'CERN effect', whereby a company's turnover increases by $4 for every $1 of CERN supply contracts they engage in.[15]

It would be impossible for me to include all the technologies that have flowed from recent developments in particle physics, but one is important to mention because it completes the set of medical diagnosis technologies. In addition to CT scanners (Chapter 1) and MRI (Chapter 11), particle physics has also had a critical role to play in developing PET (Positron Emission Tomography) scanners. Not only does PET directly use positrons (antimatter), but its detector technology comes from the development of bismuth germanium oxide (BGO) crystals used to detect particle showers. More than 1,500 PET scanners have been built with these crystals, at a cost of between $250,000 and $600,000 per scanner. During the LHC era new crystals were needed to withstand the radiation damage from the huge collision rate, which led to a new type of crystal, called the LYSO crystal. These new crystals have a faster response time and produce three times more light than BGO crystals. They are now the industry standard for PET scanners. CERN's knowledge transfer team ensured this happened even before the technology was used in the LHC, where it is only now being incorporated into detectors for the upgraded LHC programme.

Will any of the current technologies in CERN's portfolio have an impact like the web? It's impossible to tell. The LHC Computing Grid has not yet had the same impact in everyday life but is already used extensively outside of particle physics. It has provided access to more computing power than ever before in other scientific fields. Even in its earliest days the Grid allowed for the design of new anti-malarial drugs and the analysis of 140 million chemical compounds – a task that would have taken a standard computer 420 years. CERN's infrastructure and openly shared knowledge base is helping other scientists enter the realm of big data, creating entirely new ways of working in other fields.

This shift towards shared resources has now become an everyday phenomenon. Companies around the world have adopted the same approach of building large data warehouses, or clouds, where data is stored and accessed on remote servers rather than stored on your local computer. If you use cloud services like Google Docs, Dropbox or others, they are all built in a similar way. The difference between commercial cloud systems and the Grid system is where the data is stored. Grid computing means that the data storage and computing power is distributed on many different computers, instead of being stored in large corporate-owned cloud server warehouses. Today, as users become increasingly frustrated about their data being locked in with individual companies – think Microsoft's proprietary .docx or .xlxs formats or Apple's iTunes music collections – aspects of Grid technology are increasingly being adopted as solutions to challenges in cloud computing. A key goal here is *interoperability*: the ability to openly port between systems.[16] This is very much in the spirit of CERN and of Berners-Lee's creation of the web. A kind of optimum cloud-grid system may eventually help particle physicists too: it could overcome the size limits of cloud systems and enable physicists to simply use well-maintained public infrastructure.

We aren't yet done with the ways that the Large Hadron Collider has influenced the modern world because we have not discussed this experiment's largest impact of all: training highly talented people. The LHC and its detectors are international, inspirational megascience. Many of the best and brightest young minds from around the world join physics because of large projects like this, and thousands of them go on to complete PhDs in the area. It would be remiss of me not to address the question, 'What happens after that?' It might seem that their path ahead ought to be clear, but this couldn't be further from the truth.

In some areas of physics there are more than a hundred applicants for every postdoctoral job. After this, there are even fewer academic positions or permanent jobs at large laboratories. Over time, the majority of these highly trained and highly skilled people face an enormously tough decision: to stay or to leave. Deep specialisation brings unique challenges, and researchers who come to the end of short-term contracts often have no option but to move country again for the few available jobs, or change to another line of work. This might be all right for people who have the financial means to wait for the next suitable position to arise, but for many – including me – that wasn't possible.

Personally, I have faced this precipice more than once in my career. I also know from the experiences of many close friends, colleagues and peers that I was not alone in feeling an extraordinarily heavy emotional toll when forced to consider leaving behind the big-picture curiosity-driven physics research that I loved so much. Yet this also forced me to think about the many skills I have that I could apply elsewhere. I had skills in data science, in problem solving, in public speaking and in writing. I had experimental skills that could be used in industry and a demonstrated ability to see through long-term projects. I started to rejig my CV and to look on job websites. In time, I thought that I might thrive in a startup, in policy or in consultancy. I realised I could, in fact, do any of those things and enjoy them and make an impact in the world. I came to terms with the thousands of other, better-paid things I would be good at.

It's a fact that most people with PhDs in physics will end up leaving academic research. A report surveying 2,700 former CERN researchers found that 63 per cent of them are working in the private sector in areas like advanced technologies, finance and IT. The skills they bring are hugely sought-after in these sectors, skills like problem solving, programming, large-scale data analysis, science communication and international collaboration. In the UK alone, there is a shortage in so-called 'STEM' (Science, Technology, Engineering and Maths) skills to the tune of 173,000 people, despite the UK having a reputation as a global leader in science and technology.[17] The need for this kind of talent is only going to grow.

When I started looking for stories of particle physicists who had used their skills in other areas, I didn't have to go far to find some incredible innovators. Take the example of Elina Berglund, a PhD physicist involved in the hunt for the Higgs boson at CERN. She realised there was a huge knowledge gap about women's reproductive cycles, so started tracking her body data including temperature. Before long she realised that she could apply her skills in statistics and data analysis to help know when she was fertile and figured that the idea might help other women who wanted a natural way of managing their hormonal cycles. The result was the app Natural Cycles, which now has more than 1.5 million users worldwide. As of 2020 this is the only app that has FDA approval for use as a contraceptive – a life-changing technology for many women.

There is now a demonstrated path from particle physics and other curiosity-driven research fields in physics into high-tech startups, and Silicon Valley in particular. It is not just the higher pay that attracts physicists to these jobs; there is also an almost infinite variety of problems to solve once they look beyond the boundaries of their original domain of expertise. Especially in the US, the path from physics PhD programmes into Silicon Valley is now so well trodden and so well funded compared to academic research that it can be hard for physics to hang on to its best graduates.

When my own stay-or-leave moment arrived a few years after I completed my PhD, I decided that the only way I could stay in physics would be on my own terms. I couldn't change the external problems of short-term contracts, pay or funding, but I could control my own environment. I spent time building a community of like-minded physicists, especially women, so that I could see people around me who were like me and thus feel less isolated. I learned to ask for what I needed, which emboldened me to walk into the offices of leaders in my field and ask them to support my research – once, I even asked for my own job to be created. It paid off. I decided to work on things I was passionate about, things like public engagement and improving research culture, alongside my research, even if it meant pushing back against a system that said I shouldn't 'waste time' on these things. I wasn't on a one-woman mission to change physics – that would be a fool's errand – but I worked to create an environment where I felt productive, welcome and content. I knew that if it didn't work out, I would be happy to walk away.

In the end, I stayed. Together with my colleagues, I built a new lab from scratch for a small experiment which uses an ion trap to mimic particle accelerators. I was trying to understand how particle beams would behave in future colliders. I took on my first PhD student, and we commissioned the equipment. I'd never built an experiment from scratch before, and it was a huge learning curve. I couldn't believe how many things went wrong or how long some parts took to implement. One afternoon two years after I decided to stay, we were fixing some electronic noise and grounding issues when we first saw a little blip appearing above the noise on the oscilloscope screen. We were trapping and extracting ions: our first major milestone. That afternoon

I got special permission to open champagne in the lab, and we drank it from plastic cups. It wasn't exactly a Higgs boson-level achievement, but I could still barely believe it. Our experiment worked.

Looking back at this time, the thing that strikes me most is how lucky I was to have great people around me, not just during such a trying time in my career, but through the whole journey, from my earliest teachers, to mentors along the way, to the unwavering support of my PhD supervisor, my colleagues and people I later found out had been championing me without my even knowing. Physics, I have learned, is much more than the search for how the Universe and everything in it works. That's just the big question we coalesce around. Physics is all about people. It seems obvious when I say it like that, doesn't it?

Nowhere is this clearer than in the incredible story of the LHC, in which more than 10,000 scientists learned how to work together towards a common goal based on sheer curiosity. That feat alone is more than worth the investment. But of course, this story does not end with the Higgs boson. The LHC physicists are still hard at work every day – as we all are – because with new data, new ideas, and new experiments large or small we can ask new questions, getting a little closer to answers, and continuing to make progress in our quest to understand everything.

So far the LHC has produced lots of interesting, thought-provoking results, albeit not any quite as significant as finding the Higgs boson. New results are emerging every day: in the past ten years the LHC has found more than fifty new hadrons – particles composed of quarks – which are further testing our knowledge of the strong force. Some of these were even predicted in Gell-Mann's theories but were unobtainable until recently. LHC physicists have now found many particles consisting of four quarks (tetraquarks) or five quarks (pentaquarks) joined together and are still figuring out the details of how they work. Nature continues to provide new particles aplenty, but they are all covered in the Standard Model of particle physics.

While these new particles have quirks that are helping to refine the Standard Model, so far the big hopes that we would find exotic new particles in the energy range of the LHC have not been realised. In a way this is a positive thing: we are ruling out theories at a rate perhaps never seen before in the history of physics. This opens up the potential for creative new ideas and gives new areas of focus. When I ask my experimental particle physics colleagues if they are disappointed by this – because many were so hopeful of finding exotic new particles that theorists predicted lay beyond the Standard Model – most of them are surprisingly upbeat. After all, they are looking for what is real, regardless of their favourite theories. Their hard work now is sifting through the enormous amount of data produced at the LHC to see what secrets nature has in store.

Don't think, however, that we are just filling in the details, or that the journey of physics is almost complete and all the big things have been discovered. They most assuredly have not. Once again, we must look at the gaps. Despite the incredible success of the Standard Model, our equations cannot reconcile gravity with all the other forces. We don't know if there is one Universe, or whether we live in a so-called 'multiverse'. We know that neutrinos have mass and can change form, but no one knows why.[18] We don't know why we're surrounded by matter and not antimatter. We do not know the nature of the dark matter that permeates our Universe. In many ways, the Higgs boson is just the start.

13

Future Experiments

Each year around 1,500 physicists and engineers gather at the International Particle Accelerator Conference to share their work. Their projects range from proposed 100km-long colliders, to the tiniest industrial accelerators. The conference moves venue each year between Asia, the Americas and Europe, and in May 2019 it was held for the first time in Australia, in Melbourne. I had the honour of being invited to give the opening plenary talk.

Before the conference I had struggled with what to say. It wasn't the size of the audience. I'd spoken in front of larger groups and I knew how to handle those nerves. Rather, it was the expertise of the audience that put me on edge. This was by far the most important talk I'd ever been asked to give in my own field. I could follow the precedent of what others had done before and provide an expert overview of the status of our field, with plenty of technical details about particle accelerators. Yet somehow, when I started writing the talk, something quite different came out of me.

At first, I started writing just to get thoughts out of my head. I wrote not about the details of physics, but instead about the more human aspects of our field, about working together, how we got to where we are today and the lessons we have learned. I wrote about research culture and about how we can work together to address the challenges we face in the future. As my thoughts developed, I slowly realised that I wasn't going to rewrite this talk. This was a huge professional risk.

Physicists talk about physics at conferences, not about people. What if I lost the respect of my community by putting my expertise to the side and bringing this story along instead? As a newly hired faculty member I had a lot riding on it.

On the day of the presentation I took my place nervously at the front of the hall, greeted the local government minister and waited for the conference chair to introduce me. My slides were already uploaded. I closed my eyes and focused on my breathing. When the moment came, I walked up the stairs onto the stage and turned to face the audience. Under the glare of the lights I could see my collaborators from Europe, Japan, the United States and Australia, from lab directors who I knew only by reputation to colleagues with whom I'd spent midnight shifts eating pizza. Somewhere out there were also my new students at the University of Melbourne, who had never heard me speak before. I took a deep breath and began.

I told them what I had learned in this journey through twelve experiments in our field. I had been asked by the organisers to reflect on our past achievements, but also to speak about where the future might take us. So I began with my thoughts on our current position in this inspiring, big-picture, Universe-spanning journey to understand our world.

I can't help but draw parallels between where we started this journey in the late nineteenth century and where we stand in the field of particle physics as we enter the third decade of the twenty-first century. Perhaps we are on the verge of a period of transformation as enormous as that of the discovery of the nucleus, the electron and the whole subatomic and quantum world. We're looking out for a twenty-first-century version of Röntgen seeing a green flash on a screen in his lab, or Rutherford's astonishment at particles bouncing straight back from a thin gold foil. It will surely appear among complex data on a computer rather than as a flash on a screen, but the essence is the same. We're looking for something that makes us go 'hmm… that's strange'. But we cannot simply wait for these things to appear.

Discoveries have never been entirely serendipitous. People *make* discoveries: it is only by supporting those who want to get out there and build experiments to test nature that we can reach that next stage in our understanding. Luckily, this journey is already well underway. Thousands of scientists around the world – including many who were in the audience of my talk – are already planning, building and upgrading experiments small and large. Curiosity is taking them to the very edge of what is technologically possible and beyond.

Many of the proposed next-generation experiments are large collaborative ones, and for good reason. The big questions we are now asking – What is the nature of dark matter? Why is there an asymmetry between matter and antimatter in the Universe? Is there a grand unified theory that can describe everything in physics? – can't be solved by an individual or a small team working in isolation. The questions have grown too big for that. As a result, the experiments we need to answer them will almost certainly be large and complex too.

Professor Daniela Bortoletto, the Head of Particle Physics at Oxford, puts the state of the field succinctly: 'The Standard Model particles make up only about 5 per cent of the matter-energy content of the universe. The remaining 95 per cent of the universe is composed of what we do not know: dark matter and dark energy. Since we do not have any experimental evidence pointing us to the origin of the dark sector, I believe that the best way to make progress is through precision studies of the Higgs boson'.

By attempting to discover the nature of the Higgs boson, Bortoletto and her colleagues are trying to find out if the Higgs boson breaks the known laws of physics. Perhaps there are many different Higgs particles that act in strange ways. If there are, or if the Higgs decays or interacts in unexpected ways, we will have found a flaw, or a knowledge gap, at the heart of the Standard Model.

Physicists are no longer asking 'does dark matter exist?' (we think it does) but 'what is the nature of dark matter?' While progress requires both theory and experiment, dark matter presents a unique experimental challenge. There's no shortage of theories that can describe dark matter, yet the only thing we know about it for sure is that it fails to interact. We might find dark matter by seeing its failure to interact as 'missing energy' either at the LHC or in future colliders. The idea is reminiscent of the way the mystery of beta decay led us to neutrinos, but where neutrinos had a theory that helped experimentalists find them, we don't have that for dark matter: we are led by experimental data. With 95 per cent of the mass of the Universe still undetected, the stakes couldn't be higher.

Investigating these questions requires a 'Higgs factory' – a new particle collider which can produce thousands and thousands of Higgs bosons, along with the invention of a new generation of precision particle detectors, which is the aspect Bortoletto is focused on. The LHC cannot provide all the answers as to the true nature of the Higgs, so almost everyone in the field agrees that a Higgs factory ought to be a high-energy electron-positron collider with a collision energy as close to 1 TeV as possible. What is not yet agreed upon is the shape of the machine – linear or circular – or the technology on which it is based. There is likely to be only one electron-positron collider built as a Higgs factory, so we must choose which to build.

The 30km-long International Linear Collider (ILC) is now ready to be built in Japan, if governments can agree to proceed – a 'pre-lab' phase was approved in 2021. The Compact Linear Collider that CERN have been working towards for twenty years is another option.[1] The two projects already work together under the Linear Collider Collaboration, now directed by former LHC project manager Lyn Evans. Alternatively, the next big machine might be a circular accelerator 100km in circumference, under consideration at CERN (the Future Circular Collider, FCC) and in China (the Circular Electron Positron Collider, CEPC), where in addition to high-energy electron-positron collisions, the high-energy beams will blast out 50MW of unwanted synchrotron radiation – as we saw in Chapter 7 – all day, every day, as they screech around the ring. We must design and prepare these colliders now, so that one of them might be ready for when the LHC is due to finish operations in around 2036.

The Director of the John Adams Institute for Accelerator Science, Professor Philip Burrows, believes the linear version, and in particular the ILC, is the most mature design and is most likely to get us to a Higgs factory soonest. Unlike a circular design, the linear collider can also be upgraded in the future simply by extending its length. This would increase its energy reach if dark matter particles, supersymmetric particles – from a theory that predicts all matter particles have a heavier 'supersymmetric' partner – or other particles beyond the Standard Model start appearing. Bortoletto meanwhile points out that the linear collider option does not allow a later upgrade to a proton-proton collider, whereas investing in a circular tunnel means it can be reused just like the LHC re-used the LEP tunnel. The decision will come down to politics, cost and collaboration as much as physics. Whichever one is built, Bortoletto and Burrows (or perhaps their students by then!) will be there, ready to go.

Longer term, reaching ever higher energies depends on particle accelerators growing ever larger in size, despite advances in superconducting magnets and radiofrequency technology. While some researchers propose building experiments on the Moon or in space, a breakthrough in the field of plasma physics may allow us to shrink down accelerators by a factor of at least a thousand. The materials from which we make the radiofrequency cavities of particle accelerators – copper and superconducting materials – can only sustain a certain intensity of electric field before they spark or break down. This sets a physical limit to how hard we can push particles, which in turn determines the overall accelerator length. Teams of my colleagues at Oxford and Imperial College London, along with many others around the world, are instead trying to make *plasma accelerators*.

The idea is to use a high-powered laser – or even another particle beam[2] – to generate a plasma, a state of matter where atoms are already ionised. A plasma can sustain enormous electric fields, which electrons or other particles can surf along and gain energy. This has been demonstrated in the lab and has successfully accelerated particles, but it's not quite ready to roll into a particle physics experiment yet. Getting high-quality beams on demand, and controlling them, is going to take a few more years.

While it is still early days for plasma accelerators, it is definitely exciting. I always tell my students that once plasma accelerators are sufficiently advanced, I'll happily jump ship and start designing those instead. Right now, I believe they will be used in tandem with our more conventional technologies, rather than as a replacement, so I am starting to think of ways to perhaps combine the two.

Discoveries are by no means on hold while we invent these future colliders. The LHC continues to be upgraded to provide more and more data. We already know that the Standard Model is inherently flawed: it doesn't contain gravity. It can't explain why there's more matter than antimatter in the Universe. It doesn't include dark matter or dark energy. It also doesn't explain why neutrinos have mass. There *must* be more.

It would be naive to think the answers to these questions will necessarily come from particle colliders. Another field of physics may produce results that give us that next big breakthrough. Smaller experiments focused on more specific questions may get there first, and their results could be followed up at colliders. One example is the detectors being built around the world to hunt for dark matter. In Australia the first Southern Hemisphere dark matter experiment is currently under construction in the Stawell Underground Physics Laboratory (SUPL), located 1km underground in a former gold mine.

My colleagues at the International Particle Accelerator Conference knew all of these projects well. This is why I decided to talk about the way our field has not only grown our knowledge of particle physics, but also led to societal change. The twelve experiments we've seen in this book provide us with lessons about how to proceed. The stories of Brookhaven, Fermilab and CERN, and the quest for knowledge about the invisible reality of matter and forces, can give us insight on how to handle the unknowns we face in our present and future.

When I asked my colleagues what they think society can learn from the journey of particle physics, I expected a range of different answers. But I was wrong – they all said the same thing: we can learn how to collaborate. Complex undertakings like particle physics lead us to innovate, to try – as humans always will – to create order, to understand, to have knowledge and to seek wisdom. It all comes down to the compulsion we seem to have to continually step into the unknown. We strive for more, for better, and while our physical resources may be constrained, our human capacity to generate new ideas is almost limitless. It is through collaboration and new ways of working that we can realise this capacity and encourage creativity like never before.

When I think about what makes our world 'modern', I have a global sense in mind. I think about the enormous progress society has made on almost every front. New inventions have led to an increase in productivity so that goods are less scarce. Growth has led to a positive sum economy. There are more people on Earth, and they are living better lives than ever before. More people are receiving education and becoming literate. In 1930 only 30 per cent of people over fifteen years old were able to read and write now that figure is 86 per cent globally. Since 1990, every single day on average, 130,000 people have been taken out of extreme poverty, even with constantly increasing population growth. Yet despite the enormous progress in the last century, more than nine in ten people today don't think our world is getting better.[3] They could be right.

We are facing unprecedented challenges: climate change, endangered biodiversity, water scarcity, energy demands, ageing populations and, of course, pandemics and infectious diseases. With such constant newsworthy threats to our existence, no media outlet is going to remind us daily of the long-term trend that in fact humans are living longer and better lives, because we take this fact for granted. Which is odd, because there is no guarantee that we will live longer and better than our ancestors.

I am optimistic. I believe we will overcome the challenges we face as a species with innovative solutions, which is why I think it's crucial that we understand the process that creates new knowledge and ideas. If a subject as esoteric-sounding as particle physics has changed our world so profoundly, there are surely many other areas of research – not just in science but in all areas of enquiry – that we have also overlooked. This curiosity-driven research is exactly the kind of pursuit that can transform our future in ways we can't yet imagine. More than ever, this is a time for us to cultivate the skills to face unknowns and work together for the good of all humankind.

Looking back through these twelve experiments, I see that in addition to learning how to collaborate, there are three key ingredients we need in order to face the challenges of the future: the ability to ask good questions; a culture of curiosity; and the freedom to persist. We need to ask just the right question, in the right context and at the right time. The key is to ask questions that leave us open to the idea that we might be wrong, setting aside our own biases. No matter how well an idea serves us, our questions have to be framed in such a way that we can change our minds. J. J. Thomson did not ask 'does the electron exist?' and then conclude 'no' when his first experiments showed some results that were inconsistent with his hypothesis. A *good* question must dig into the heart of an unknown. Good questions like 'what is the true nature of cathode rays?' tend to create lots of smaller questions in their wake, like 'do cathode rays bend in an electric field?' Asking these smaller questions is essential. In fact, it was these smaller questions that led Thomson to find inconsistencies that revealed a path forward. Only when all the smaller questions were addressed could he find the answer to his big question. The result was the discovery of the electron.

Also important is that we don't necessarily have to answer all the questions we ask. Some of them may not be answered for centuries. Yet good questions are powerful motivators: for all the wondrous achievements we've made trying to understand the nature of matter and forces, it is not the answers that keep us going. It is the questions.

The environment in which we ask these questions is no less important. We've seen how curiosity can lead to remarkable breakthroughs, but what does a culture that supports curiosity look like? It looks like the brainstorming sessions a friend of mine runs in which adding to an idea is allowed, but there is no critiquing. It looks like the time one of my PhD students was inspired by a YouTube video about designing rollercoasters using Artificial Intelligence to do the same with particle accelerators and I wholeheartedly encouraged him. Nascent ideas need nurturing at first. Wilson wasn't trying to invent a particle detector, Röntgen wasn't trying to invent a medical technology and CERN wasn't on a mission to invent the World Wide Web. It was only because they worked in a culture that supported human curiosity that their ideas flourished.

Enabling this culture is *hard*. What with our goals and objectives and plans and shareholders and reports and deadlines, who has time for that? But it is worth it. The act of seeking knowledge reveals more spectacular landscapes when we do not have a destination in mind.

Finally, we need to grant ourselves – by which I mean individuals, teams, society, humanity – the freedom to persist in our endeavours. It wasn't possible for us to gradually piece together the Standard Model of particle physics without many false starts, a lot of confusion and the frustratingly slow accumulation of knowledge over decades. Doing anything for the first time is incredibly difficult. But doing something for the first time when only a handful of other humans understand it is even harder. When I say we need to cultivate the freedom to persist, I am talking about more than willpower and doggedness. I am talking of tangible things like time, space and resources.

We need to encourage environments in which people can follow their curiosity, take intellectual risks and flourish. We are at a point of great opportunity. If we can learn to value the creative nature of science, to nurture curiosity and promote both intellectual depth and breadth within ourselves and in young people around us, I have no doubt we can face what lies ahead. Yet in one critical respect we are not doing so.

Throughout this book we've seen example after example of the ways in which our understanding of some of the most fundamental parts of physics have led to tangible outcomes. It would be easy, knowing this, to decide to fund research based on its potential for so-called impact. Many governments have chosen to do so, at least in part, but they are often focused on quick results. In a field like particle physics, which has had such an enormous influence on society in ways that couldn't be predicted and over timescales that are unfathomable to most politicians, what can one say to such short-term thinking?

If focusing on quick results were the model, Rutherford's labs would never have existed and Robert Wilson's proposal to build the Tevatron would never have passed Congress. Peter Higgs himself once said that in the current academic system, he wouldn't even have a job[4] since he simply didn't churn out enough papers. In today's system he would be doubly excluded since he also had no claim to short-term real-world impact. Today we expect hyper-productivity, accountability and value for money. Though it feels uncouth to talk of curiosity and money almost in the same breath, if we want to make big breakthroughs in the future we will need money.

Creating the freedom to persist requires us to acknowledge the role that curiosity-driven research plays in our society. This is a profound shift in the way we think about the value of science. In fact, I would extend that argument to the way we think about the value of research in general. Humans are – as Hannah Arendt says – question-asking beings. As we've repeatedly seen, the person who makes a discovery is possibly the worst person to ask what it will be used for, or what might come from it. We must agree to support question-asking and curiosity because it adds to our flourishing as humans, not because it might put our country in a better economic position or manage to eke out another half a per cent efficiency from solar panels – although it may do these things anyway. Let us not miss out on discoveries because we weren't able to see their value before they were made.

What's more, we need to learn how to do this collectively. There is nothing more powerful than humans who come together in collaborative endeavour. The handful of experiments in the later stages of this story could never have come about if it weren't for individuals taking risks and working together over decades. Imagine how different our lives would look now if previous generations had not done so?

When I stood up to speak that day in Melbourne, I touched on many of these points. To be honest I don't remember walking off the stage, or what the next few speakers had to say. When the coffee break arrived, I tried to get to the coffee stand but every few metres someone would appear, beaming, expounding on various elements of my talk. Some of them I knew, most of them I didn't. The chair of the speaker committee found me later that day, and happily reflected on the many conversations my talk had spurred in our community.

Through years of research I had learned something much larger than just physics. I had learned how to follow my curiosity and step into the unknown, to ask good questions and to persist through the many barriers that have impeded me. When I reflected that learning back to my community, I found that others in my field were there with me, cheering me on. It was something I didn't even realise I had been missing: a sense of belonging.

Exhausted by the conference, I decided to leave the crowd behind on the last Friday evening and headed to my department. The silence of the basement greeted me, and my steps resonated down the concrete corridor as I walked past a painted mural of the Big Bang and along to a set of wooden doors. I swiped my brand new university ID across the card reader, pushed the door open and walked in.

Past the scattered desks and cardboard boxes, I stood within the confines of my new lab. The walls are made of concrete blocks thick enough to shield the outside world from the particle beams that used to be – and will be again – created here. I have big goals for this lab. I picture the space anew: white walls, flashing safety lights, yellow hazard signs, black cables and copper accelerating structures. I see students, staff and collaborators – my tribe – busy at work.

The questions I'm asking now explore the physics of particle accelerators, where the dual needs of physicists on the one hand and of medicine and industry on the other send me back to the drawing-board to the point where physics and invention collide. My mind is full of questions about the physics of particle beams and their swirling, non-linear dance of oscillations and electromagnetic interactions. Later, we will get to questions of engineering, of cost and implementation, but for now my curiosity starts with the interactions of the tiny, invisible world of particles and connecting that with ideas which could make our lives better, even if in the far distant future. This is my small niche within an enormous spectrum of physics, within science, within the human journey of research and knowledge.

Experiments designed to understand matter and forces stretch back hundreds of years and our current understanding relies on thousands, possibly tens of thousands, of experiments. In this journey we have followed but a handful of them. These experiments have moulded the way we think about the nature of the Universe, created many of the technologies we use every day and co-created our modern, interconnected, world.

Here in my new lab I stand on the verge of a conversation with the unknown, grateful to have the time and space in which that conversation can take place. I know that this lab is where failure and frustration will sit alongside success. It will take an outpouring of energy, curiosity and creativity to transform this space from a shell of concrete blocks into a source of new knowledge, but I also know that I wouldn't want to put that energy into anything else.

I cannot promise that we will change the world, but at least we know how to proceed. One experiment at a time.

ACKNOWLEDGEMENTS

Although my name is on the cover, this book – like the experiments within it – was made possible by many others. My heartfelt thanks go to:

The many colleagues, researchers and experts who agreed to be interviewed, showed me around labs or helped me with this story: Rob Appleby, Elisabetta Barberio, Alan Barr, Daniella Bortoletto, Phillip Burrows, Harry Cliff, Frank Close, Sonia Contera, Les Gamel, Rob George, David Jamieson, Sneha Malde, Steve Myers, John Patterson, Larry Pinsky, Harry Quiney, Sergey Romanov, Werner Ruhm, Martin Seviour, Marco Schippers, Ian Shipsey, Geoff Taylor and Rachel Webster. A special mention to Ray Volkas, for being my go-to theoretical physicist and supporter. I should be quite clear, however, that any physics mistakes are entirely my own;

Many mentors who have coaxed my journey in physics into existence: Roger Rassool, who helped me believe that I could pursue this subject; and Ken Peach, my DPhil supervisor, for always supporting my science communication alongside my research, and for continually seeing slightly more in me than I saw in myself;

Chris Wellbelove, my agent, for your creativity, persistence and patience. I couldn't have chosen a more insightful guide for my journey to becoming a writer;

My wonderful editors, Alexis and Edward, and the teams at Bloomsbury and Knopf. Your vision for this book has helped me grow in every way. Thank you for helping me tell this story;

The fabulous Oxford Writing Circle for creating a space for me to share – with hands shaking – the very first words of this story, and the London Writers Salon for very many hours 'alone but together' in our sacred virtual writing space. You are my not-so-secret productivity hack;

My many colleagues, collaborators, staff and students for being so understanding when I repeatedly disappeared for long stretches to write this book. Special thanks go to my brilliant research students, who are a constant source of inspiration. I can't wait to get back in the lab, or as Lucy would say: 'Let's physics';

Alex de H, Ian R, Jan M and Sarah R for your unwavering friendship throughout this process. Kippy and Ross, my parents, for placing such radical importance on education. And my grandmother Enid, who at 100 years old has lived through much of this story. Jason and Grace, for your belief and support. Finally, to my identical twin sister, Megan. There are no words to express my gratitude: you are truly the most incredible woman in my life and always will be.

NOTES

INTRODUCTION

1   You may note I haven't included gravity here, despite us experiencing it on a daily basis. Gravity is not included in the Standard Model, and is *unbelievably* weak compared to the other three forces. The question of why this is, and how to integrate these theories together, constitutes one of the grand challenges of physics in the twenty-first century.

1   CATHODE RAY TUBE: X-RAYS AND THE ELECTRON

1   Usually called a cathode ray tube. Technically, I describe here a Crookes–Hittorf tube, but all these tubes were similar. These experiments must be done in a near-vacuum or the cathode rays will collide with gas molecules and get scattered or lost. The average distance between collisions is called the 'mean free path' and applies to all molecules, atoms and other particles travelling through a gas. The mean free path of a cathode ray in air is minuscule, so the tubes only work under vacuum.

2   *Nobel Lectures, Physics 1901–1921*, Elsevier, Amsterdam, 1967.

3   See Otto Glasser, *Wilhelm Röntgen and the Early History of the Roentgen Rays*, Norman Publishing, San Francisco, 1993. One wonders if the distant part of his family tree that had found fame making elaborate pieces of furniture with quirky mechanical features had been an influence. For more on this, see Wolfram Koeppe, *Extravagant Inventions: The Princely Furniture of the Roentgens*, Yale University Press, New Haven CT, 2012.

4   Glasser, *Röntgen*.

5   One of them was given to him by a Ukrainian physicist named Ivan Puluj, who had reported in 1889 that photographic plates became black when

exposed to cathode rays. Röntgen and Puluj had worked together in Strasbourg, and Röntgen regularly attended the lectures given by Puluj, who developed a special cathode ray tube called a 'Puluj lamp' that was mass-produced for a time. Puluj used it to produce images of the skeletons of a mouse and a foetus. So, did Puluj actually beat Röntgen to the discovery of X-rays? Probably not, because what Puluj did not realise was that he was seeing fundamentally different rays from the ones inside the tube, and that was the key insight that led to Röntgen being credited with the discovery.

6   Glasser, *Röntgen*.

7   At first the name 'Röntgen rays' stuck, particularly in Germany, but it wasn't to be in the rest of the world, where with time the catchier 'X-rays' persisted. His name is instead memorialised in a unit of radiation known as the 'Röntgen', although in some medical departments you'll find a 'Röntgenology' department instead of the X-ray department.

8   At this point in time there was a geographical divide in thinking on the nature of cathode rays. Most German scientists thought they were a kind of light, while most British scientists leaned instead towards them being made of a type of particle.

9   Lord Rayleigh (J. W. Strutt), *The Life of Sir J. J. Thomson OM*, Cambridge University Press, Cambridge, 1943, p. 9.

10   J. J. Thomson, 'XL. Cathode Rays', *Philosophical Magazine Series 5*, vol. 44, 1897, pp. 293–316.

11   J. J. Thomson, *Recollections and Reflections*, G. Bell, London, 1936.

12   Thomson, 'XL. Cathode Rays'.

13   Albemarle Street in London, where the Royal Institution is located, was made the first one-way street in the world to manage the many carriages of visitors for these Friday Evening Discourses.

14   The existence of the electron was accepted by both the 'particle' and 'light/aether' schools of thought because it was already identified as the unit of electricity, and in the latter theory was supposed to be a disturbance in the aether.

15   *Proceedings of the Royal Institution of Great Britain*, vol. 35, 1951, p. 251.

16   Lewis H. Latimer, 'Process of manufacturing carbons', *US Patent* 252,386, filed 19 February 1881.

17   P. A. Redhead, 'The birth of electronics: Thermionic emission and vacuum', *Journal of Vacuum Science and Technology*, vol. 16, 1998.

18   In the meantime, he'd designed and built the radio transmitter that sent the first radio transmissions across the Atlantic in December 1901.

As per his agreement with the Marconi company, the credit went to Marconi, despite the invention being Fleming's. He later felt that Marconi had been ungenerous towards him.

19 The company had formed after a legal battle around the design of the light bulb. They produced bulbs entirely of Swan's design except for the filament.

20 The triode was invented by Lee De Forest in 1906 for a device called the *audion*, an early audio amplifier. See Lee De Forest, 'The Audion: A new receiver for wireless telegraphy', *Transactions of the American Institute of Electrical and Electronic Engineers*, vol. 25, 1906, pp. 735–63.

21 As early as the 1920s, a few people had the idea of moving the X-ray source and detector and taking lots of X-rays from different angles. The objects in the middle of the setup would be in focus, and the stuff on the outside would be blurred out and could be ignored. This idea was called 'tomography' and around ten different people, all working independently, produced a series of patents on the idea between 1921 and 1934. They could all fairly lay claim to its invention, but the first real working version was by Gustave Grossman from Germany in the late 1930s, through his company Siemens-Reiniger-Veifa GmbH. But the method remained clunky and difficult to use and still wasn't good at showing the density differences between different types of tissue inside the body.

22 No ordinary cow brains would do, either. The method of killing in abattoirs at the time damaged them, so he had to travel to Jewish kosher houses where the cow brains were less damaged and suitable for his CT experiments.

23 S. Bates et al., *Godfrey Hounsfield: Intuitive Genius of CT*, British Institute of Radiology, London, 2012.

24 Ibid.

2 THE GOLD FOIL EXPERIMENT: THE STRUCTURE OF THE ATOM

1 After his studies in Oxford, Soddy had emigrated to Canada hoping to gain a professorship in Toronto but was unsuccessful, so ended up employed as a demonstrator in chemistry at McGill University.

2 Muriel Howorth, *Pioneer Research on the Atom: The Life Story of Frederick Soddy*, New World Publications, London, 1958.

3 Ibid.

4   Richard P. Brennan, *Heisenberg Probably Slept Here: The Lives, Times and Ideas of the Great Physicists of the 20th Century*, J. Wiley, Hoboken NJ, 1997.

5   Named after Marie Curie's home country, Poland.

6   Ernest Rutherford, 'Uranium radiation and the electrical conduction produced by it', *Philosophical Magazine*, vol. 57, 1899, pp. 109–63.

7   M. F. Rayner-Canham and G. W. Rayner-Canham, *Harriet Brooks: Pioneer Nuclear Scientist*, McGill-Queen's University Press, Montreal, 1992.

8   T. J. Trenn, *The Self Splitting Atom: A History of the Rutherford–Soddy Collaboration*, Taylor and Francis, London, 1977.

9   Howorth, *Pioneer Research*.

10 A. S. Eve, *Rutherford: Being the Life and Letters of the Rt. Hon. Lord Rutherford*, Cambridge University Press, Cambridge, 1939, p. 88.

11 Ibid.

12 Ibid., p. 118.

13 This tradition known as the marriage 'bar' applied to women in most professions and only ended in the 1950s in Canada, while it lingered in the United States and other Western countries until as late as the mid-1970s.

14 Rayner-Canham, *Harriet Brooks*.

15 John Campbell, *Rutherford: Scientist Supreme*, AAS Publications, Washington DC, 1999.

16 To this day, physicists regularly admit they mentally associate atoms and particles with little coloured balls, a picture so incorrect that many wish they had never been taught it when they were young.

17 H. Nagaoka, 'Kinetics of a system of particles illustrating the line and the band spectrum and the phenomena of radioactivity', *Philosophical Magazine*, vol. 7(41), 1904.

18 C. A. Fleming, 'Ernest Marsden 1889–1970', *Biographical Memoirs of Fellows of the Royal Society*, vol. 17, 1971, pp. 462–96.

19 Arthur Eddington, *The Nature of the Physical World*, Macmillan, London, 1928.

20 United States Environmental Protection Agency, 'Mail irradiation'. Available online at https://www.epa.gov/radtown/mail-irradiation. Accessed 29 March 2021.

21 P. E. Damon et al., 'Radiocarbon dating the Shroud of Turin', *Nature*, vol. 337, 1989, pp. 611–15. https://doi.org/10.1038/337611a0.

22 C. J. Bae, K. Doouka and M. D. Petraglia, 'On the origin of modern humans: Asian perspectives', *Science*, vol. 358 6368, 2017. 10.1126/science.aai9067.

23 Sarah Zielinski, 'Showing their age: Dating the fossils and artifacts that mark the great human migration', *Smithsonian Magazine*, July 2008. Available online at https://www.smithsonianmag.com/history/showing-their-age-62874/. Accessed 29 March 2021.

24 C. Buizert et al., 'Radiometric 81Kr dating identifies 120,000-year-old ice at Taylor Glacier, Antarctica', *Proceedings of the National Academy of Sciences*, vol. 111, 2014, pp. 6,876–81. https://doi.org/10.1073/pnas.1320329111.

25 The asteroid hypothesis was originally put forward by physicist Luis Alvarez (see Chapter 8) and his son. There has since been a debate about whether it was volcanism, rather than an asteroid, but in 2020, modelling of each scenario has re-centred the asteroid model as most likely. See Chiarenza et al., 'Asteroid impact and not volcanism caused the end-Cretaceous dinosaur extinction event', *Proceedings of the National Academy of Sciences*, vol. 117, 2020, pp. 17,084–93. https://doi.org/10.1073/pnas.2006087117

26 Adam C. Maloof et al., 'Possible animal-body fossils in pre-Marinoan limestones from South Australia', *Nature Geoscience*, 3, 2010, pp. 653–9. https://doi.org/10.1038%2Fngeo934.

3 THE PHOTOELECTRIC EFFECT: THE LIGHT QUANTUM

1 At the time the word 'corpuscle' was used, making Newton a protagonist of 'corpuscularianism'.

2 The idea of the aether persisted until the nineteenth century when, in 1887, the Michelson–Morley experiment showed that the luminiferous aether did not exist, causing sufficient embarrassment to physics to open the way for Einstein's special relativity to be accepted.

3 Diffraction does happen in the situation of the fence, but the light that passes through the missing slats dominates compared to the minor effect of diffraction. The effect increases when the 'slit' is similar to the wavelength of the light, which is only a few hundred nanometres.

4 Or if you don't have time – or your pond is frozen – beautiful versions are available online including this version from Veritasium. Available online at https://www.youtube.com/watch?v=Iuv6hY6zsdo.

5 At the time they described the effect in terms of electric charge, as electrons weren't discovered for another decade.

6 Lenard won the Nobel Prize in 1905 for his contributions to the photoelectric effect. He was publicly antisemitic and described Einstein's work in relativity as 'Jewish fraud'. He later became Chief of 'Aryan Physics' under Hitler.

7   B. R. Wheaton, 'Philipp Lenard and the Photoelectric Effect, 1889–1911' *Historical Studies in the Physical Sciences*, vol. 9, 1978, pp. 299–322.

8   The vacuum level he achieved was a millionth of a millimetre of mercury, roughly 10^{-6} millibars in today's units, which is well into the modern 'high vacuum' range, an astounding feat with glass tubes!

9   J. J. Thomson, *Conduction of Electricity Through Gases*, Cambridge University Press, Cambridge, 1903.

10   R. A. Millikan and G. Winchester, 'The influence of temperature upon photo-electric effects in a very high vacuum', *Philosophical Magazine*, vol. 14, 1907, pp. 188–210. https://doi.org/10.1080/14786440709463670.

11   Ibid.

12   R. A. Millikan, *The Autobiography of R. A. Millikan*, Prentice-Hall, Inc., Englewood Cliffs, 1950.

13   There is some evidence that Einstein's main works are at least partly Mileva's work. See Pauline Gagnon, 'The forgotten life of Einstein's first wife' *Scientific American*, 2016. Available online at https://blogs.scientificamerican.com/guest-blog/the-forgotten-life-of-einsteins-first-wife/

14   G. Holton, 'Of love, physics and other passions: The letters of Albert and Mileva' (part 2), *Physics Today*, vol. 47, 1994, p. 37.

15   Astrophysical objects can radiate in X-rays. X-ray astronomy is just one additional way we have of viewing the Universe in a different frequency range which goes from radio-astronomy all the way through to gamma rays. Google some images from the Chandra X-ray observatory – they are spectacular!

16   This is a theoretical object, although it can be approximated in experiments.

17   We're a lot less sensitive to violet light than to blue, so even if the violet part of the spectrum were brighter, our eye would perceive blue as brightest.

18   There's a story that physicists often tell that Planck started looking at blackbody radiation because he was asked by the German authorities to calculate how to make light bulbs more efficient, but there's little truth to the rumour.

19   Planck considered that the light given off by the blackbody must be due to the vibrations of so-called *resonators* – oscillating electric charges – which produce electromagnetic radiation. Each resonator could vibrate

at any frequency. This came from the statistical mechanics view of thermal physics.

20   For a lengthy but insightful explanation that works through Planck's pivotal paper see A. P. Lightman, *The Discoveries: Great Breakthroughs in Twentieth-Century Science*, Pantheon, New York, 2005.

21   *h* has a value of 6.626×10^{-34} Joule-seconds (or in SI units $m^2 kg/s$). The key point is that it's a very small number. The unit of Joule-second is the unit of action – energy multiplied by time – which represents the energy of a wave, divided by the frequency of the wave in Hz (s^{-1}).

22   A. Hermann, *The Genesis of Quantum Theory*, MIT Press, Cambridge MA, 1971.

23   Helge Kragh, 'Max Planck: The reluctant revolutionary', *Physics World*, vol. 13(12), 2000. Available online at https://doi.org/10.1080/14786440709463670.

24   Abraham Pais, *Subtle is the Lord: The Science and Life of Albert Einstein*, Oxford University Press, Oxford, 2005, p. 382.

25   By 1909–10 he had dropped the photoelectric effect temporarily to embark on a major series of experiments which would also make his name. His idea was quite ingenious, and if you've studied any physics, it's why Millikan's name rings a bell. Millikan knew that electrons were particles after J. J. Thomson's work in 1897, but he came up with the most precise way to measure the electric charge of the electron. This work started to close out any doubt that the electron was also the thing that travelled though wires as electricity. University students often recreate this famed 'oil drop' experiment even today. But for me, it is his lesser known and much more arduous experiments on the photoelectric effect that are most impressive.

26   Recently, Caltech has removed Millikan's name from its buildings, honours and assets because of an idea he seemingly struggled to accept, that of race equality. 'Millikan lent his name and his prestige to a morally reprehensible eugenics movement that already had been discredited scientifically during his time'. See https://www.caltech.edu/about/news/caltech-to-remove-the-names-of-robert-a-millikan-and-five-other-eugenics-proponents.

27   *Light emitting diodes (LEDs)* just use the reverse process, using electricity to produce light.

28   This method is called *photoplethysmography* (PPG) and is also used in
      pulse oximeters.

29   These orbitals are technically the place where the electron is found
      about 90 per cent of the time: the electron's position is uncertain, due to
      Heisenberg's uncertainty principle.

30   V. Kandinsky, "Reminiscences" (1913), in V. Kandinsky, Kandinsky:
      Complete Writings on Art. Edited by Kenneth C. Lindsay and Peter
      Vergo. 2 vols. Boston: G. K. Hall and Co.; London: Faber and Faber,
      1982, pp. 370.

31   What does it mean for humans to have a wavelength? Not much really;
      any object with mass and energy can be said to have a wavelength,
      and once we're at walking speed ours is something like 10^{-37}m which is
      smaller than we can ever hope to measure. Sorry to disappoint.

32   This is described online in https://medium.com/the-physics-arxiv-blog/
      physicists-smash-record-for-wave-particle-duality-462c39db8e7b which
      refers to Sandra Eibenberger et al., 'Matter-wave interference with
      particles selected from a molecular library with masses exceeding 10000
      amu', *Physical Chemistry Chemical Physics*, vol. 15, 2013, pp. 14,696–700.
      https://doi.org/10.1039/C3CP51500A.

33   A. Tonomura et al., 'Demonstration of single-electron buildup of an
      interference pattern', *American Journal of Physics*, vol. 57, 1989. https://
      doi.org/10.1007/s00016-011-0079-0.

34   R. Rosa, 'The Merli–Missiroli–Pozzi two-slit electron-interference
      experiment', *Physics in Perspective*, vol. 14, 2012, pp. 178–95.
      https://doi.org/10.1119/1.16104.

35   For a detailed but accessible explanation of energy bands in solids, see
      Chad Orzel, 'Why do solids have energy bands?', *Forbes*, 2015. Available
      online at https://www.forbes.com/sites/chadorzel/2015/07/13/why-do-
      solids-have-energy-bands/#2acb0b9d1080.

4 CLOUD CHAMBERS: COSMIC RAYS AND A SHOWER OF NEW PARTICLES

1   Franz Linke took twelve balloon flights during his PhD studies. Alfred
      Gockel and Karl Bergwitz also used hot-air balloons before Victor Hess.

2   Not to be confused with the cosmic microwave background radiation,
      discovered by Penzias and Wilson in 1965, which is the faint
      electromagnetic radiation left over from an early stage of formation of
      the Universe.

3   C. T. R. Wilson, 'XI. Condensation of water vapour in the presence of dust-free air and other gases', *Philosophical Transactions of the Royal Society of London, Series A*, vol. 189, 1897, pp. 265–307. https://doi.org/10.1098/rsta.1897.0011.

4   C. T. R. Wilson, 'On the ionization of atmospheric air', *Proceedings of the Royal Society*, vol. 68, 1901. https://doi.org/10.1098/rspl.1901.0032.

5   Sue Bowler, 'C.T. R. Wilson, A great Scottish physicist: His life, work and legacy' (conference paper), *Royal Society of Edinburgh*, 2012.

6   In a modern glassblowing lab, the glass used is Pyrex – the same stuff you have in your kitchen – which is widely available and made to a standard recipe, so today's scientific glassblowers, people with rare technical skill, can join pieces originating from Japan, the United States or Europe without a hitch. But Wilson would have used soda glass, which was far more fragile and difficult to shape.

7   C. T. R Wilson, Nobel Lecture, 12 December 1927.

8   Ibid.

9   G. Zatsepin and G. Khristiansen, 'Dmitri V. Skobeltsyn', *Physics Today*, vol. 45(5), 1992. https://doi.org/10.1098/rspl.1901.0032.

10 Harriet Lyle, interview with Carl Anderson, January 1979. Available at http://resolver.caltech.edu/CaltechOH:OH_Anderson_C. Accessed 6 April 2021.

11 C. D. Anderson, 'The Positive Electron', *Physical Review*, vol. 43, 1933, p. 491. https://doi.org/10.1103/PhysRev.43.491.

12 Dirac didn't invent the term 'antimatter', which was coined by Arthur Schuster in 1898 (A. Schuster, 'Potential matter: A holiday dream', *Nature*, vol. 58, 1898). However, his idea was purely speculative and invoked antigravity, while the modern idea of antimatter does not.

13 John Hendry, *Cambridge Physics in the Thirties*, Adam Hilger, London, 1984.

14 E. Cowan, 'The picture that was not reversed', *Engineering and Science*, vol. 46(2), 1982, pp. 6–28. Available online at: https://resolver.caltech. edu/CaltechES:46.2.Cowan. Last accessed 18 January 2022.

15 Werner Heisenberg, letter to Wolfgang Pauli, 31 July 1928. In W. Pauli, *Scientific Correspondence*, vol. 1, Springer Verlag, Berlin, 1979.

16 A. Pais, *Inward Bound*, Oxford University Press, Oxford, 1986, p. 352.

17 Blackett had to wait until 1948, after he had made another major discovery, while Occhialini would miss out.

18 These are truly fundamental particles so they are not composed of electrons, and there are also two ghostly particles called neutrinos

18   Harry E. Gove, *From Hiroshima to the Iceman: The Development and Applications of Accelerator Mass Spectroscopy*, Institute of Physics Publishing, Bristol, 1999.

19   There had been prior work using cyclotrons in the United States. See Richard Muller, 'Radioisotope Dating with a Cyclotron', *Science*, vol. 196, 1977, pp. 489–94, although this paper did not demonstrate carbon dating with AMS. Gove's team was the first to do this and to use a tandem accelerator for the experiment, still the technology of choice today.

20   It is unclear if Gove and Bennett ever tested the violin, but it's likely they would have found an ambiguous date, even if it were real. The period in which Stradivarius violins were made happens to correspond to a thirty-year period of low solar activity called the Maunder minimum. The lower level of solar UV in this period allowed more ozone to build up in Earth's atmosphere and as a result, cosmic rays created an excess of radioactive carbon-14 in the atmosphere. Any wood from the period of the Maunder minimum has extra carbon-14 and this results in it looking as if it might be much more recent, in this case from the 1950s. It's not possible to get rid of this date ambiguity without a more traditional historical investigation. Periods like this are known and calibrated using modern carbon dating techniques.

21   Available online at https://inis.iaea.org/search/search.aspx?orig_q=RN:47061416.

6   CYCLOTRON: ARTIFICIAL PRODUCTION OF RADIOACTIVITY

1   A version of the periodic table at this time can be found at https://www.meta-synthesis.com/webbook/35_pt/pt_database.php?PT_id=1017.

2   Herbert Childs, *An American Genius: The Life of Ernest Orlando Lawrence, Father of the Cyclotron*, E. P. Dutton, Boston, 1968.

3   Inside a conductor like a metal tube, any voltage on the outside doesn't penetrate inside.

4   Childs, *An American Genius*.

5   Ibid.

6   M. S. Livingston and E. M. McMillan, 'History of the Cyclotron', *Physics Today*, vol. 12(10), 1959. https://doi.org/10.1063/1.3060517.

7   L. Alvarez, *Ernest Orlando Lawrence: A Biographical Memoir*, National Academy of Sciences, Washington DC, 1970.

8   Childs, *An American Genius*.

9   E. O. Lawrence, 'Radioactive sodium produced by deuton bombardment', *Physical Review*, vol. 46, 1934, p. 746. https://doi.org/10.1103/PhysRev.46.746.

10　The ethics of conducting animal experiments has also undergone a major overhaul since those days.

11　Usage of cobalt-60 in the medical field has recently been identified as a nuclear proliferation risk and poses a hazard of unwanted exposure to radiation if sources are incorrectly disposed of. Usage is expected to decrease in future as accelerators are now the preferred source, as discussed in Chapter 10.

12　Today the existence of technetium in the spectrum of stars has provided evidence that heavy elements are created in stars, a process known as stellar nucleosynthesis, but this wasn't known until the late 1950s.

13　D. C. Hoffman, A. Ghiorso and G. T. Seaborg, 'Chapter 1.2: Early Days at the Berkeley Radiation Laboratory' in *The Transuranium People: The Inside Story*, University of California, Berkeley and Lawrence Berkeley National Laboratory, 2000.

7　SYNCHROTRON RADIATION: AN UNEXPECTED LIGHT EMERGES

1　Available online at https://www.nytimes.com/1998/06/09/nyregion/commemorating-a-discovery-in-radio-astronomy.html.

2　Bell-Burnell is today a well-known professor and advocate for diversity in science. In 2018 she donated her $3 million winnings from the Breakthrough Prize to fund scholarships to increase diversity in physics.

3　The University of Illinois physics department created a competition to name it, and among the entries was an 84-character-long German name which thankfully didn't win: Ausserordentlichehochgeschwindigkeitselektronenentwickelndesschwerarbeitsbeigollitron, which translates to 'Hardworking by golly machine for generating extraordinarily high velocity electrons'.

4　Rolf Wideröe had also come up with a similar concept.

5　The vacuum vessel really is referred to by experts as a 'doughnut'.

6　Even today we still call the characteristic oscillations of particles in accelerators *betatron oscillations*.

7　Theoretical physicists had figured this out as early as 1897 and it was used in early attempts to describe how electrons were expected to lose energy and spiral into the nucleus in Rutherford's model of the atom. Most physicists had stopped looking at it, however, after Bohr explained the atom's stability using quantum mechanics, making the problem irrelevant to understanding the atom. In the 1940s with the advent of the betatron, electrons were being pushed up to energies sufficient for this effect to be a problem.

8   A few years later, Ed McMillan published a full theory of the synchrotron, and despite being aware of Oliphant's idea, he did not include a single mention or citation of Oliphant's work.

9   This surfer in a headwind analogy was provided by my former RAL colleague, Dr Stephen Brooks, now at Brookhaven National Laboratory.

10   The magnet was made of lots of flat pieces or 'laminations' joined together, like a metal cake sliced into tiny thin pieces. This method of segmenting a magnet helps when you're ramping the magnet up and down quickly as it means that electric 'eddy' currents don't flow wildly throughout the steel.

11   They had been beaten a month earlier by two British physicists, F. K. Goward and D. E. Barnes, who converted a small betatron at Woolwich Arsenal into an 8MeV electron synchrotron.

12   H. Pollock, 'The discovery of synchrotron radiation', *American Journal of Physics*, vol. 51, 1983. https://doi.org/10.1119/1.13289.

13   You can tell if your sunglasses are polarised by holding them at 90 degrees to a known polarised pair in a shop and looking through the lenses: if it is completely dark they are polarised. Rotate them back to 0 degrees and they will let light through again.

14   M. L. Perlman et al., 'Synchrotron radiation: Light fantastic', *Physics Today*, vol. 27, 1974. https://doi.org/10.1063/1.3128691.

15   They were awarded the Nobel Prize for this work just two years later in 1915.

16   The electron beam has a limited 'lifetime' due to many effects that gradually cause electrons to be lost: scattering of the electrons from the tiny amount of gas left in the vacuum, scattering of the electrons with each other and quantum excitations. Remember, electrons emit light in quanta – photons – with each electron emitting around 100 photons per turn in the synchrotron. This quantised effect causes sudden 'bumps' to affect the electrons and causes the beam to diffuse, limiting its lifetime in the ring.

17   The emitted radiation power is proportional to the particle mass to the fourth power, i.e. mass squared and squared again.

8   PARTICLE PHYSICS GOES LARGE: THE STRANGE RESONANCES

1   Alvarez was accompanied by two other physicists, Harold Agnew and Lawrence Johnson.

2   https://www.manhattanprojectvoices.org/oral-histories/carl-d-andersons-interview.

3   Along with her nephew, Otto Frisch.

4   https://www.manhattanprojectvoices.org/oral-histories/evelyne-litzs-interview.

5   Winston S. Churchill, *Victory*, Rosetta Books, New York, 2013.

6   Alvarez was the person who informed Ernest Lawrence about the breakthrough.

7   It would be another eight years before physicists identified that both of these particles were actually versions of what we now call the kaon.

8   R. Armenteros et al., 'LVI. The properties of charged V-particles', *The London, Edinburgh, and Dublin Philosophical Magazine and Journal of Science*, vol. 43, 1952, pp. 597–611. https://doi.org/10.1080/14786440608520216.

9   There are three types of pion: positive, negative and neutral.

10 This was a formidable challenge since protons, as mentioned earlier, are around 2,000 times heavier than electrons and need stronger magnets to contain them at high speed, but their hands were forced because electron beams lost much of their energy to synchrotron radiation, as we saw in the last chapter.

11 Luis W. Alvarez, *Alvarez: Adventures of a Physicist*, Basic Books, New York, 1987.

12 Eric Vettel, *Donald Glaser: An Oral History*, University of California, Berkeley, 2006.

13 Ibid.

14 There is a popular myth among physicists that Glaser invented the bubble chamber while staring into a glass of beer. Unfortunately it's not true. The actual story is that at one point Glaser found himself at a local pub called the Pretzel Bell with a gathering of his colleagues, who started teasing him about his experiments. One of them pointed to the bubbles in beer and said 'Gee, Glaser, it should be easy, you can see tracks damn near anywhere!' The pub later put his photograph on the wall and claimed he'd made his invention in the pub, and the myth was perpetuated from there.

15 Vettel, *Glaser*.

16 L. M. Brown, M. Dresden and L. Hoddeson, *Pions to Quarks: Particle Physics in the 1950s*, Cambridge University Press, Cambridge, 1989, p. 299.

17 Vettel, *Glaser*.

18 Women had been analysing particle data since at least the 1920s, including the ones recording scintillations in Vienna mentioned in Chapter 5. In the 1940s women were also employed to analyse nuclear emulsion data, particularly in Cecil Powell's laboratory in Bristol. During the war, as men

went off to fight, many women found new work opportunities, including as 'computers', doing detailed calculations to solve the differential equations that arose in many projects. When the first electronic calculating machines came along, the machines took the title 'computer' and the women naturally took over the role of computer programmers instead, although they were often given no acknowledgement and their stories were for a long time ignored. So when it came time to analyse particle tracks from bubble chambers, the division of labour along stereotypical gender lines was barely in question. It was clearly 'women's work'.

19 M. Gell-Mann, 'Isotopic spin and new unstable particles', *Physical Review*, vol. 92, 1953, p. 833. https://doi.org/10.1103/PhysRev.92.833 and M. Gell-Mann, 'The interpretation of the new particles as displaced charged multiplets', *Il Nuovo Cimento*, vol. 4(S2), 1956, pp. 848–66. https://doi.org/10.1007/BF02748000.

20 'Alternating gradient' was a new focusing concept which came from a breakthrough in accelerator physics made at the Cosmotron, where it was found that by reversing the polarity of every other magnet a beam can be confined to a much narrower beam pipe (not one you can drive a car through!). This enabled machines to be built with smaller and cheaper magnets, while the beams could reach even higher energies.

21 The idea of quarks was invented independently by George Zweig, who called them 'aces'.

22 V. E. Barnes et al., 'Observation of a Hyperon with Strangeness Minus Three', *Physical Review Letters*, vol. 12(8), 1964, pp. 204–6. https://doi.org/10.1103/PhysRevLett.12.204.

23 Vettel, *Glaser*.

24 Mary Palevsky, *Atomic Fragments – A Daughter's Questions*, University of California Press, 2000, p. 128.

25 This is one of the areas of my own research and Wilson is, admittedly, a bit of a hero of mine. My own PhD thesis was on the design of new accelerators, similar to cyclotrons, which we designed specifically to enhance modern particle therapy. These machines are called 'Fixed Field Alternating-gradient' or FFA accelerators and were first invented in the 1950s and 1960s in the United States by the MURA collaboration.

26 Named after Australian-British physicist William Henry Bragg who first predicted it way back in 1904.

27 U. Amaldi, 'History of hadrontherapy in the world and Italian developments', *Rivista Medica*, vol. 14(1), 2008.

28   L. Hoddeson, 'Establishing KEK in Japan and Fermilab in the US: Internationalism, nationalism and high energy accelerators', *Social Studies in Science*, vol. 13(1), 1983. https://doi.org/10.1177/030631283013001003.

9   MEGA-DETECTORS: FINDING THE ELUSIVE NEUTRINO

1   From Fred Reines' Nobel Lecture, available online at https://www.nobelprize.org/uploads/2018/06/reines-lecture.pdf.

2   Technically, this process captures the neutrino's antimatter twin, the antineutrino, but Reines and Cowan didn't know that, and it makes little difference to us for the time being. This is the opposite of radioactive beta decay from Meitner and Hahn's early experiments and is called *inverse beta decay*.

3   The invention of these tubes is difficult to pin down but is thought to have happened in either Russia or the United States. They are now primarily developed and sold by a Japanese company, Hamamatsu.

4   These would have been cathode ray oscilloscopes. See Chapter 1.

5   This idea of a *whole-body radiation counter* was later used in medicine for understanding the degree to which radioactive material, whether natural or human-made, is taken up, circulated and used by the human body.

6   Technically this is an antineutrino, the neutrino's antimatter equivalent, which is required as the decay has to conserve 'lepton number': the electron has lepton number +1 and the antineutrino has −1, balancing each other out.

7   This doesn't happen easily because you can't just directly fuse protons together. First, four protons have to fuse into two deuterons, then you add additional protons to get two helium-3 nuclei and finally you fuse those together to form a single helium-4. Along the different steps of the chain, neutrinos, gamma rays and positrons are all released.

8   The gamma rays, on the other hand, would bounce around in electromagnetic interactions, take hundreds of thousands of years to reach the surface and eventually come out as visible light.

9   Whether we detect the matter or antimatter version of a new particle first is of little consequence to physicists nowadays. But neutrinos are interesting in this sense. We define neutrinos and antineutrinos from the way they interact with other particles, by the values of so-called lepton number, but it is still not known whether they are in fact distinct or whether neutrinos are their own antiparticle.

10   Pontecorvo's life is a fascinating story; see Frank Close's books *Neutrino*, OUP, Oxford, 2010 and *Half Life*, Oneworld, London, 2015.

11   You can take a virtual tour here: https://www.snolab.ca/facility/vr-tour

12   Kamioka Nucleon Decay Experiment was originally designed to measure the decay of a proton. It was able to set limits on the proton lifetime but turned out to also be great for neutrinos and was upgraded over time for this purpose. Neutrino interaction in the water can produce electrons or positrons moving faster than the speed of light (that is, the speed of light in water, which is slower than in vacuum), an effect equivalent to a sonic boom of a jet. This produces a cone of light which we call Cherenkov radiation, which is what Super-Kamiokande was designed to measure.

13   https://www.symmetrymagazine.org/article/june-2013/cinderellas-convertible-carriage.

14   https://www.techexplorist.com/scientists-measured-neutrinos-originating-interior-earth/29364/.

15   https://www.sheffield.ac.uk/news/nr/nuclear-particle-physics-research-study-watchman-uk-us-boulby-1.828008.

16   https://www.popsci.com/science/article/2012-03/first-time-neutrinos-send-message-through-bedrock/.

17   C. Thome et al., 'The REPAIR project: Examining the biological impacts of sub-background radiation exposure within SNOLAB, a deep underground laboratory', *Radiation Research*, vol. 88(4.2), 2017, pp. 470–4. doi: 10.1667/RR14654.1.

10 LINEAR ACCELERATORS: THE DISCOVERY OF QUARKS

1   Typically operating at around 20–50MHz.

2   The klystron created milliwatt-level power in the GHz frequency range. Engineers and physicists in the accelerator field tend to refer to both 'microwave' (over 1GHz) and 'radio' (MHz to GHz) frequencies as 'radiofrequency'.

3   James P. Baxter, *Scientists Against Time*, Atlantic-Little Brown, Boston, 1947, p. 142.

4   This was called the 'Tizard Mission', after British chemist Henry Tizard, who instigated it.

5   Unbeknown to them, Yoji Ito had also independently developed a magnetron in Japan in 1939.

6   Not to be confused with Lawrence's Rad Lab at Berkeley; the MIT Rad Lab was given this intentionally confusing name so that people thought they were simply working on basic physics, not military-related radar work.

7 Frank J. Taylor, 'The Klystron Boys: Radio's Miracle Makers', *Saturday Evening Post*, 8 February 1942, p. 16.

8 The brothers later took the company public, in 1956.

9 John Edwards, 'Russell and Sigurd Varian: Inventing the klystron and saving civilization', available at https://www.electronicdesign.com/technologies/communications/article/21795573/russell-and-sigurd-varian-inventing-the-klystron-and-saving-civilization. Accessed online 29 June 2021.

10 Christophe Lécuyer, *Making Silicon Valley: Innovation and the Growth of High Tech, 1930–1970*, MIT Press, Cambridge MA, 2006.

11 E. Ginzton, 'An informal history of SLAC: Early accelerator work at Stanford', *SLAC Beam Line*, special issue 2, 1983.

12 The calculation was based on the assumption of needing to resolve down to about 1 per cent of the radius of the proton or neutron, which gives an electron energy of 20GeV using the de Broglie wavelength of electrons.

13 In the quark model, all the heavy and strange particles could be described as having combinations of quarks and their antimatter versions, antiquarks. It meant that a proton would have two up quarks and one down quark, while a neutron would have one up quark and two down quarks. The *mesons*, like the pion and kaon, were composed of two quarks, or a quark together with an antiquark. The strange particles, known as *baryons*, had three types of quark or antiquark. At some point along the way, all the particles that interact via the strong force were given the name *hadrons*.

14 There was a small difference from the gold foil experiment, though. The electrons in this case would lose some energy in the collision, so it was referred to as an *inelastic* collision, as opposed to the *elastic* collisions of the gold foil experiment. The experiment is known as *deep inelastic scattering*.

15 Until in 1999 another physics project, the gravitational wave interferometer (LIGO) took the title.

16 Today it has been upgraded to 50GeV.

17 Michael Riordan, 'The discovery of quarks', *Science*, vol. 256, pp. 1,287–93. https://doi.org/10.1126/science.256.5061.1287.

18 Available online at https://hueuni.edu.vn/portal/en/index.php/News/the-road-to-the-nobel-prize.html. Accessed 5 October 2020.

19 Remember the hydrogen atom has just a proton and an electron, while deuterium has a proton and a neutron.

20 For their discovery Friedman, Kendall and Taylor were later awarded the 1990 Nobel Prize in physics.

21 Ionising radiation was used to treat skin lesions all the way back in 1897, after the discovery of X-rays, but with only low energies available, the rays couldn't penetrate into the body, so were no good for most tumours. It wasn't until accelerators came on the scene providing electron energies (and thus X-ray energies) in the 'megavoltage' range that the beams had enough penetrating power.

22 Before the LINAC there had been a few competing technologies for radiotherapy, from Van de Graaff and Cockcroft-Walton machines to betatrons. Most were either too big or didn't provide a high enough dose rate for quality therapy. In the UK a 3m-long, 8MeV X-ray machine was the first LINAC in the world to treat patients, but it was a large device that couldn't be moved around to deliver beam from many angles. A few smaller 4MeV machines were also built in the UK at this time and, before long, machines had been installed in Australia, New Zealand, Japan and Russia. The new Varian machines quickly replaced these early installations as radiotherapy grew.

23 See https://www.iceccancer.org/cern-courier-article-developing-medical-linacs-challenging-regions/.

24 See https://www.computerworld.com/article/3173166/bill-nye-backed-startup-uses-particle-accelerator-to-make-solar-panels-60-cheaper.html.

25 Very High Energy Electron (VHEE) therapy, combined with fast dose delivery to achieve the so-called 'FLASH' effect, is a hot topic in the field at the minute.

11 THE TEVATRON: A THIRD GENERATION OF MATTER

1 'R. R. Wilson's congressional testimony, 1969', *Fermilab*. Available online at https://history.fnal.gov/historical/people/wilson_testimony.html. Accessed 31 May 2021.

2 Ibid.

3 Wilson styled it after the cathedral at Beauvaix, in France.

4 https://history.fnal.gov/goldenbooks/gb_wilson.html

5 L. Hoddeson, A. W. Kolb and C. Westfall, *Fermilab: Physics, the Frontier, and Megascience*, University of Chicago Press, 2009, p. 101.

6 Reports are mixed on whether the ferret cleaning method worked but at any rate a robotic system later replaced Felicia.

7   Called 'slow resonant extraction'.

8   Hoddeson, Kolb and Westfall, *Fermilab*.

9   The 'J/Ψ' particle is a combination – or 'bound state' – of a charm and anti-charm quark, as quarks cannot be 'found' in isolation. Burton Richter's group at SLAC named the new particle the 'Ψ' while Samuel Ting's group working at Brookhaven named it the 'J', and the two were so close in time that the particle became known by a combination of both, the J/Ψ.

10 D. C. Hom, L. M. Lederman et al., 'Observation of high mass dilepton pairs in hadron collision at 400 GeV', *Physical Review Letters* vol. 36(21), 1976, p. 1,236. https://doi.org/10.1103/PhysRevLett.36.1236.

11 Technically, it means that if the new particle doesn't exist, the chance of the data looking the way it does is one in 3.5 million. It's a bit annoying to phrase it like that, but there's no way around that conditional 'if' statement because that's how the stats work. But I prefer to phrase it as 'there's less than a one in 3.5 million chance that the data is a fluke'.

12 'Revisiting the b revolution', CERN Courier, 2017. Available online at https://cerncourier.com/a/revisiting-the-b-revolution/.

13 These first superconducting materials were niobium-zirconium and niobium-titanium.

14 J. Jackson, 'Down to the Wire', *SLAC Beam Line*, vol. 73(9), spring 1993, p. 14.

15 Over at Brookhaven this lesson was learned the hard way, where they were developing magnets for a slightly smaller superconducting accelerator called ISABELLE. Brookhaven made fewer prototypes and then got a shock when the magnet which worked in a short length failed to work properly at full scale, a problem that played a large role in the cancellation of the ISABELLE project in 1982.

16 The first idea for a collider appears to have started with Rolf Wideröe in 1953. Later ideas came from Novosibirsk in Russia, from the Midwestern Universities Research Association (MURA), from Brookhaven, SLAC and the Cambridge Electron Accelerator (CEA).

17 1.6TeV collisions came first on 13 October 1985.

18 This decay also produces a W boson, the electrically charged carrier of the weak force.

19 See https://cerncourier.com/a/superconductors-and-particle-physics-entwined/.

20 See https://www.elekta.com/radiotherapy/treatment-delivery-systems/unity/.

21 See http://bccresearch.blogspot.com/2010/09/global-market-for-mri-systems-to-grow.html.

22 Judy Jackson, 'Down to the wire', *SLAC Beam Line*, vol. 23(1), 1993, p. 14. https://www.slac.stanford.edu/history/newsblq.shtml

12 THE LARGE HADRON COLLIDER: THE HIGGS BOSON AND BEYOND

1 Evans is also the only current Fellow of the Royal Society in my field of accelerator physics.

2 M. Krause, *CERN: How We Found the Higgs Boson*, World Scientific, Singapore, pp. 98–107.

3 Interview with Lyn Evans, *BBC Wales* (archived). Available online at https://www.bbc.co.uk/wales/scifiles/interviewsub/liveevans.shtml.

4 Krause, *CERN*.

5 W and Z bosons weigh in at around 70GeV.

6 This number seems quite specific, because it is based on the mass of the Higgs boson, which we're getting to, hold on!

7 Given by physicist David Miller of University College London.

8 J. D. Shiers, *Data Management at CERN: current status and future trends,* Proceedings of IEEE 14th Symposium on Mass Storage Systems, 1995, pp. 174–81, doi: 10.1109/MASS.1995.528227.

9 See https://www.bondcap.com/pdf/Internet_Trends_2019.pdf.

10 The per capita GDP is $1,369 and the population of Kashmir is 12.55 million, so the total GDP is around $17 billion. See https://thediplomat.com/2020/08/perpetual-silence-kashmirs-economy-slumps-under-lockdown/.

11 'Tevatron experiments close in on Higgs particle', *Symmetry*, July 2011. Available online at https://www.symmetrymagazine.org/breaking/2011/07/27/tevatron-experiments-close-in-on-higgs-particle.

12 Gianotti would later become the first female Director General of CERN, a role she holds at the time of writing.

13 For details, see https://seeiist.eu.

14 For information on CERN's technology portfolio, see https://kt.cern/technologies.

15 See *The Impact of CERN*, CERN-Brochure-2016-005-Eng, December 2016. https://home.cern/sites/home.web.cern.ch/files/2018-07/CERN-Brochure-2016-005-Eng.pdf. Accessed 11/10/2021. Also see P. Castelnovo et al., 'The economic impact of technological procurement

for large scale infrastructures: Evidence from the Large Hadron Collider', *CERN* paper, 2018. Available online at https://cds.cern.ch/record/2632 083/files/CERN-ACC-2018-0022.pdf.

16 David Villegas et al., 'The role of grid computing technologies in cloud computing', in B. Fuhrt (ed.), *Handbook of Cloud Computing*, Springer Verlag, Berlin, 2010, pp. 183–218.

17 P. Amison and N. Brown, *Evaluation of the Benefits that the UK has derived from CERN*, Technopolis Group, 2019.

18 By the way, the neutrino masses don't account for the missing mass of the Universe.

13 FUTURE EXPERIMENTS

1 This project is based on high-frequency 'X-band' accelerating technology: one of their test systems is – at the time of writing – being installed in my new laboratory at the University of Melbourne and will grow our programme in medical and industrial accelerators as well as contributing to future linear collider developments.

2 At CERN, Edda Gschwendtner is overseeing AWAKE (the Advanced Wakefield Experiment) taking proton beams at 400GeV and using them to make a plasma channel to accelerate electrons.

3 This survey was conducted in 2015, before the COVID pandemic, Trump's presidency and Brexit. Available online at https://ourworldindata.org/a-history-of-global-living-conditions-in-5-charts.

4 Available online at https://www.theguardian.com/science/2013/dec/06/peter-higgs-boson-academic-system.

INDEX

Accélérateur Grand Louvre d'analyse élémentaire, 116–17

Accelerator Mass Spectrometry (AMS), 115

Adams, Frank Dawson, 34, 43

adenosine triphosphate (ATP), 149

Agnew, Harold, 292

AIDS, 152

Akeley, Lewis, 121–2

Alexander, Frances Elizabeth, 140

Allibone, Thomas, 101, 103, 107

alpha radiation, 31–3, 37–40, 42, 71, 75, 78–9, 84, 97–8, 100–6, 109–12, 123, 130, 161, 181–2, 209

Alternating Gradient Synchrotron (AGS), 174, 294

Alvarez, Luis, 91, 159–63, 165–7, 169, 171–3, 205, 283, 292–3

Alzheimer's disease, 238

American Physical Society, 171, 237

americium, 42, 135

Anderson, Carl, 79–85, 128, 161

angular momentum, 164, 174

anode ray tubes, 108

anthrax, 42

antimatter, 81–3, 119, 166, 190, 198, 244, 258, 263, 267, 270, 287, 295, 297

Apeizon, 108

Arendt, Hannah, 69, 273

argon, 137, 190

astatine, 120, 135

asteroids, 43, 283

Atkinson Morley Hospital, 24

atomic clocks, 66

atomic nucleus
discovery of, 40–1, 43–4, 209
and splitting the atom, 96–118
and strong force, 211–12
structure (protons and neutrons), 112, 118

AWAKE experiment, 301

bacteria, 63

Bahcall, John, 190

Bailey, Melanie, 117

bananas, 42

baryons, 174, 297

Becquerel, Henri, 30–1

beer, 170, 291

Bell-Burnell, Jocelyn, 140, 291

Ben Nevis, 75–6

Bennett, Charles, 95–6, 113–15, 290

benzene ring, 149

Berglund, Elina, 260

Bergwitz, Karl, 286

berkelium, 135

Berners-Lee, Tim, 252–3, 259

beryllium, 90, 108–10, 132

beta decay, 168, 180–3, 188–90, 194, 220, 267, 295

beta radiation, 31–3, 42, 71, 75, 78–9, 101, 140, 161

betatrons, 140–4, 146–7, 291–2, 298
betatron oscillations, 291

Bethe, Hans, 182

Bevatron, 165–7, 171–3, 177, 205

Big Bang, 250, 274

bismuth germanium oxide (BGO) crystals, 258

Bjorken, James, 210
black holes, 88, 140, 147, 244
blackbody radiation, 52–3, 61, 284
Blackett, Patrick, 81–3
Blau, Marietta, 86–8
Blewett, John, 142–3, 145
Blumer, Molly, 127
Bohr, Niels, 61–2, 82, 104, 161, 182, 291
Boot, Harry, 201
borehole logging, 42
boron, 115
Bortoletto, Daniela, 267–9
Bose, D. M., 87
Bragg, Lawrence, 149
Bragg, William Henry, 57, 79, 149, 294
'Bragg peak', 177
brain scanners, 24–5
breast cancer, 177
bremsstrahlung (braking radiation), 20
Brennan, Richard P., 30
British Association for the Advancement
     of Science, 35, 41
Brobeck, William, 165
Brookhaven National Laboratory, 7, 165–6,
     169, 172–4, 190, 224, 239, 270, 299
Brooks, Harriet, 32, 35, 42
Brooks, Stephen, 292
bubble chambers, 170–2, 174–5, 183,
     185, 208–9, 238, 293–4
Burch, Bill, 108
Burrows, Philip, 268–9
Byron, Lord, 9

californium, 135
calutrons, 160
Cambridge Electron Accelerator
     (CEA), 299
Campbell, Eleanor, 150–3
canal ray tubes, 108
carbon-14, 89, 95–6, 113–14, 136
carbon ions, 177
Cassiopeia (constellation), 140
cathode rays, 11–12, 16–19

Cavendish Laboratory, 17, 20, 32, 76–8,
     81, 96, 98–101, 105–7, 112, 116,
     128, 130
centre-of-mass energy, 228
cerebrospinal fluid (CSF), 23–4
CERN, 91, 179, 188, 208, 211, 215, 224,
     227, 229, 231–2, 238, 240–60, 268,
     270, 272, 301
     see also Large Hadron Collider
Chadwick, James, 99–100, 109–11,
     128–9, 137, 161, 181, 189
Chandra X-ray observatory, 284
'chart of nuclides', 136
chocolate, 153
Chodorow, Marvin, 205
Chowdhuri, Bibha, 86–8
Churchill, Winston, 162
Circular Electron Positron Collider
     (CEPC), 268
climate change, 89–90, 237, 271
Clinton, Bill, 241
cloud chambers, 71–2, 77–86, 88, 92, 106,
     122, 128, 130, 163, 169, 185
cloud formation, 76–7
cloud services, 258–9
cobalt-60, 134, 170, 291
Cockcroft, John, 101–3, 105–12, 116–20,
     122, 127–8, 131, 141, 161, 201–2,
     205, 289
Cockcroft Institute, 113, 289
Cockcroft-Walton generators, 112–13,
     222, 298
Collier, Paul, 247
collimation system, 213–14
Compact Linear Collider, 268
condensation nuclei, 76–7
confinement, 211
Conover, Emily, 193
consciousness, 63
conservation of electrical charge, 164
conservation of momentum, 182, 188
contraceptives, 260
Coolidge, William, 142

corona effect, 107, 289
cosmic microwave background
     radiation, 286
cosmic rays, 75, 79–80, 83–90, 92–3, 95,
     98, 112, 122, 138, 163, 165–7, 169,
     171, 185, 187, 192, 197, 244, 290
Cosmotron, 165–7, 172, 174, 205, 294
Coulomb barrier, 102–3, 105
COVID-19, 149–52, 301
Cowan, Clyde, 183–90, 295
Crab nebula, 147
Crocker laboratory, 136
Crookes, William, 30
Crookes–Hittorf tube, 279
CT scanners, 23–5, 177, 238, 258
Curie, Irène, 100, 109, 129–31
Curie, Marie, 31–2, 35, 73, 100, 121, 129
Curie, Pierre, 31
curium, 135
cyclotron, invention of, 122–31
Cygnus (constellation), 140

Daresbury Synchrotron Radiation Source
     (SRS), 148, 153
dark matter, 267–8, 270
Davis, Ray, 189–92
Davisson, Clinton, 63
de Broglie, Louis, 61, 63, 179
de Broglie wavelength, 63, 65, 297
De Forest, Lee, 281
Dead Sea Scrolls, 43
DeBeers, 116
Debye, Peter, 182
decompression sickness, 137
decuplet, 174
delta particles, 174
deuterium (heavy hydrogen), 128–9,
     210, 297
deuterons, 128–30, 140, 177, 295
Diamond synchrotron light source, 152
diethyl ether, 170–1
diffraction, 46, 149, 283
dinosaurs, 43

Dirac, Paul, 81–3, 287
Dirac equation, 81, 83
DNA, 149
doping, 66, 116
double slit experiments, 45–6, 55–8,
     60–1, 63–4
drugs, 117

e/m ratio, 18
Earth, age of, 34–5
earthquakes, 194
ebola, 152
Eddington, Arthur, 40
Edison, Thomas, 21–3
Edison effect, 21–2
Edlefsen, Niels, 125–6
Edwards, Helen, 222–3, 227, 231–2, 239
Egyptian mummies, 114
Eibenberger, Sandra, 63
Eiffel Tower, 74
Eightfold Way, 174, 208
Einstein, Albert, 41, 51–2, 54–8, 61–2, 66,
     81, 83, 86, 129, 138, 161, 283–4
     $E=mc^2$, 111–12, 164
     theory of special relativity, 58, 81, 138
einsteinium, 135
electromagnetic force, 193, 195, 206,
     211, 250
electromagnetic waves, discovery
     of, 17, 48
electromagnetism, 4, 12, 47, 52, 168
electrometers, 31, 77
electron microscopes, 64–6
electron-positron colliders, 216–17,
     250, 268
electrons
     compared with muons, 91
     discovery of, 3–4, 19–23, 29–31, 36,
         38–9, 50, 280, 285
     electron neutrinos, 192–3, 224,
         233, 249
     interference patterns, 63–4
     quantised energy values, 61, 65–6

and structure of atoms, 36, 38–40, 47
    see also beta radiation
electron-Volts, 105
electroscopes, 17, 72–4, 77
electroweak force, 250
EMI, 24
epidemiology, 237
eugenics, 285
Evans, Lyndon (Lyn), 243–4, 246, 248, 268
Exner, Franz, 86

Federal Telegraph Company, 127
Fermi, Enrico, 168, 182, 184
Fermi Space Telescope, 88
Fermilab, 7, 193, 196, 220–40, 243, 249, 251, 254–5, 270
    Colliding Detector at Fermilab (CDF), 233–4, 240
    Underground Parameters Committee, 231
    see also Tevatron
fermium, 135
Feynman, Richard, 145, 207, 210
5-sigma evidence, 225–6
Fleming, John Ambrose, 22–3, 281
Fleming valves, 20, 22–3
foot-and-mouth disease, 153
fossils, 43
francium, 120, 135
Franklin, Rosalind, 149
Friedman, Jerome, 209–12, 298
Frisch, Otto, 161, 293
Fukushima Daichi reactor, 92
Future Circular Collider (FCC), 268

Gage, Matilda J., 88
gallium, 116, 120
'Gamma knife', 214
gamma radiation, 31–2, 42, 52–3, 71, 75, 83, 108–10, 131, 134–5, 170, 181–2, 184–6, 284, 295
Gamow, George, 103–5, 122, 128

Geiger, Hans, 37–40, 42, 97, 99, 209
Geiger counters, 38, 40, 82, 90–1, 130, 186
Gell-Mann, Murray, 173–4, 208, 262
General Electric, 142–5, 147, 153, 202–3
George, E. P., 90
Germer, Lester, 63
Geuricke, Otto von, 26
Gianotti, Fabiola, 255
giant magnetoresistance (GMR), 153
Ginzton, Ed, 205–6, 208, 213
Gladisch, Ellen, 86
Glaser, Donald, 168–72, 183, 293
Glasgow Royal Infirmary, 15
glassblowing, 77–8, 106, 287
'glories', 76
gluons, 211, 250
Gockel, Alfred, 286
gold foil experiment, 36–40, 47, 297
Gonzales, Angela, 221
Gove, Harry, 95–6, 113–15, 290
GPS navigation systems, 66
graphene, 149
graphite, 113, 149
gravity, 2, 12, 168, 193, 219, 249–50, 263, 270, 279
    antigravity, 287
Great Depression, 80, 133
Greenstein, Jesse, 140
Greinacher, Heinrich, 289
Griffith Observatory, 71–2
Grossman, Gustave, 281
Gschwendtner, Edda, 301
gunshot residues, 117

Haber, Floyd, 145
hadrons, 262, 297
haemoglobin, 149
Hafelekar research station, 86
Hahn, Otto, 161
Hallwachs, Wilhelm, 48
Hansen, Bill, 200–2, 204–6, 208–9, 212

Harari, Haim, 227
heavy water, 192
Heisenberg, Werner, 82
Heisenberg's uncertainty principle,
    82, 173
heliosphere, 89–90
helium ions, 66, 177
helium nuclei, *see* alpha particles
Herr Auge, 186
Hertz, Heinrich, 17–18, 30, 48, 121
Hess, Victor, 74–5, 83, 89, 286
Heuer, Rolf-Dieter, 256
Higgs, Peter, 250, 256, 273
Higgs boson, 239, 241–2, 244, 250–2,
    254–6, 260, 262–3, 267–8
Higgs field, 250–1
Hiroshima, 159–60, 162
Hodgkin, Dorothy, 149
Homo sapiens, 43
hot-air balloons, 7, 73–4, 77, 89, 286
Hounsfield, Godfrey, 24–6
Huygens, Christiaan, 45

ice cores, 43, 90
Incandela, Joseph, 255
insulin, 149
interference, 46, 58, 63–4
Intermagnetics General Corporation
    (IGC), 238
International Atomic Energy Agency
    (IAEA), 215
International Conference in High Energy
    Physics (ICHEP), 255
International Linear Collider (ILC), 268
International Particle Accelerator
    Conference, 265–6, 270, 274
International Space Station, 240
inverse square law, 72
iodine-131, 134
iron-59, 133
ISABELLE accelerator, 299
Ito, Yoji, 296
Ivanenko, Dmitri, 142–5

J/Ψ particle, 224, 299
Jansky, Karl, 139
Jefferson, Thomas, 117
John Adams Institute for Accelerator
    Science, 268, 289
Johnson, Lawrence, 292
Joliot, Frédéric, 100, 109, 130–1

Kajita, Takaaki, 189, 191–3
Kandinsky, Wassily, 62
kaons, 163, 174–5, 208, 233, 293, 297
Kapitza, Peter, 101
Kaplan, Henry, 213
Kelvin, Lord, 33–5
Kendall, Henry, 209–12, 298
Kerst, Donald, 142, 147, 153
klystrons, 202–5, 207, 209, 212, 214,
    216, 296
Kobayashi, Makoto, 227
Kornberg, Roger, 149
Koshiba, Masatoshi, 189
krypton, 137

lambda particles, 168, 175
Langmuir, Robert, 146
Langsdorf, Alexander, 72
Large Electron Positron Collider (LEP),
    250–2, 269
Large Hadron Collider (LHC), 196,
    242–56, 258–9, 262–3, 267–70
  ALICE experiment, 244
  ATLAS experiment, 244–8, 255
  CMS experiment, 244, 248, 255
  LHCb experiment, 244
  Worldwide LHC Computing Grid
      (WLCG), 254, 258–9
Latimer, Lewis, 21
Lauritsen, Charles, 109
Lawrence, Ernest Orlando, 120–33,
    136–7, 141, 143–4, 153, 160–1, 163,
    165–6, 175, 178, 293
Lawrence, John, 124, 131–3, 137,
    176–7

Lederman, Leon, 193, 224–8, 231–2, 239–40
Leksell, Lars, 214
Lenard, Philipp, 30, 48–9, 283
Leon Mow Dark Sky Site, 1–2, 4
lepton number, 293
leptons, 224–5, 227, 249
leukaemia, 132, 152
Libby, Willard, 89
light bulbs, 21–2, 147, 281, 284
light-emitting diodes (LEDs), 59, 285
linear accelerators, invention of, 205–16, 298
Linear Collider Collaboration, 268
Linze, Franz, 286
liquid hydrogen, 171
liquid scintillator, 185–6
Litherland, Albert, 96
lithium, 90, 97, 105, 108, 110–12, 117, 122, 128
Litton Industries, 202–4
Litz, Evelyne, 162
Livingston, Milton Stanley, 126–8, 130
Lofgren, Ed, 165
London Olympics, 253
Lonsdale, Kathleen, 149
Los Alamos, 161–2, 183–4, 187, 236
Lovelace, Ada, 9
LYSO crystal, 258

McDonald, Arthur, 189, 191–3
McMillan, Edwin, 143–4, 163, 292
macromolecular crystallography, 150
Maglev transportation, 239
Magnetic Corporation of America (MCA), 238
magnetic resonance imaging (MRI), 238–9, 258
magnetic spectrometer, 210
magnetrons, 201–5, 212, 214, 216
Manchester Museum, 43
Manhattan Project, 159–62, 175, 183, 220
Marconi, Guglielmo, 22, 281

Marić, Mileva, 51–2, 284
Marsden, Ernest, 37–40, 42, 97, 99, 209
Marsh, Robert, 239
Mary Rose, 153
Maskawa, Toshihide, 227
Matilda Effect, 88
Maunder minimum, 290
Maxwell, James Clerk, 47
Meitner, Lise, 161
Mendeleev, Dmitri, 119–20, 135
mendelevium, 135
mercury pumps, 50, 284
mesons, 85, 87, 174, 223, 232, 288, 297
Metropolitan Vickers (Metrovick), 102, 106–7
Meyer, Stephan, 86
Michelson–Morley experiment, 283
microwaves, 201, 203
Midwestern Universities Research Association (MURA), 299
Milky Way, 2, 140, 147
Millikan, Robert, 49–51, 55–60, 62, 66, 79–80, 124, 285
MINERvA experiment, 196
MIT Rad Lab, 202, 296
modulators, 22
molybdenum, 134
Monte Carlo method, 235–7
Moon, the, 216, 249, 254, 269
Mount Asama, 91
Mount Etna, 92
Mount Vesuvius, 92
mRNA, 149
Muographix collaboration, 288
muons, 84–5, 87–93, 98, 119, 163–4, 167–8, 192–3, 207, 217, 224–5, 227, 233, 244–5, 249, 288
muon neutrinos, 192–3, 224, 233, 249
MX-2 beamline, 150–2
myoglobin, 149

Nagaoka, Hantaro, 36
Nagasaki, 160–2

nanotechnology, 65
Napoleon Bonaparte, 117
Natural Cycles app, 260
Nature, 111, 182
Ne'eman, Yuval, 174
Neddermeyer, Seth, 83–4, 161
Neumann, John von, 235
neurotoxins, 238
neutrino geophysics, 194
neutrinos, 181–98, 207, 211, 227, 250,
    267–8, 270, 287–8, 295–6, 301
  antineutrinos, 190, 194–5, 198, 295
  discovery of, 181–9
  electron neutrinos, 192–3, 224, 233, 249
  handedness, 194
  muon neutrinos, 192–3, 224, 233, 249
  oscillation, 192–3, 263
  solar neutrinos, 189–92, 194, 196
  tau neutrinos, 233, 241, 249
neutron stars, 88, 147
neutrons
  diffusion of, 236
  discovery of, 97–8, 109–11, 128–9, 189
  sub-structure of quarks, 174–5, 209–12
Newton, Isaac, 45
Niedergesass, Felix, 107
niobium-titanium, 229
Nishijima, Kzuhiko, 173
nitrogen-15, 33
Nobel Prizes, 35–6, 83, 87–8, 98, 118,
    131, 135–6, 140, 145, 148–9, 161,
    189, 191–3
nobelium, 135
non-destructive testing, 26
nuclear emulsions, 85–8, 163
nuclear fission, 161, 184, 195
nuclear fusion, 195–6
nuclear power, 41, 118
nuclear weapons, 41, 136, 159–62, 175,
    179, 184

Occhialini, Giuseppe, 81–3, 87
Oganessian, Yuri, 136

oil drop experiment, 285
Oliphant, Mark, 36, 143, 153, 161, 292
omega minus particle, 174
Oppenheimer, Robert, 161
optical fibres, 60
orbital cutters, 257
Orr, Rich, 232
oxygen-15, 33

Pacheco, Clare, 116–17
Pacini, Domenico, 74
pair production, 83
Panofsky, W. K. 'Pief', 206
particle therapy, 176–80
partons, 210
Pastore, Senator John, 219–20
Paul Scherrer Institute, 178
Pauli, Wolfgang, 82, 182, 188–9
pedoscope, 15
Peierls, Rudolf, 182
penicillin, 149
Perey, Marguerite, 135
periodic table, 117, 119–20, 134–6,
    181, 290
Perrier, Carlo, 134
Perrin, Jean, 30
phase stability, 144, 163
Philips, 112
phosphorus, 116, 132
photodiodes, 59–60
photoelectric effect, 48–52, 55–60, 65–7,
    79, 185, 195, 283, 285
photomultiplier tubes, 185–6, 192, 295
photons, 55, 58–9, 61, 64, 83, 119, 176,
    187, 194, 206–7, 211, 250, 292
photoplethysmography (PPG), 286
photosynthesis, 89
photovoltaic cells, 59
physics, classical v. modern, 47
Pike's Peak, Colorado, 83–4
pions, 87, 163–4, 167–8, 174, 193, 206,
    208, 212, 233, 293, 297
pituitary gland, 177

Planck, Max, 52–5, 61–2, 161, 284
Planck's constant (*h*), 54, 57–8, 61
plasma accelerators, 269
plutonium, 135
pneumoencephalography, 23–4
polarised light, 146
Pollock, Herb, 142
polonium, 31–2, 73, 98, 109
Pomeranchuk, Isaak, 142–5
Pontecorvo, Bruno, 190–1
Positron Emission Tomography (PET), 90, 258
positrons, 81–4, 88–90, 98, 118–19, 128, 163, 167, 184, 188, 295–6
  see also electron-positron colliders
potassium-40, 194
Powell, Cecil, 87, 293
power stations, 92
Pozzi, Giulio, 64
p-p chain reaction, 189
pressure cookers, 169–70
promethium, 120, 135
Protein Data Bank, 150
proton accelerators, 163, 196, 205–13, 206, 208–9
proton beam, complex formations of, 247
proton therapy, 176–8, 213
protons
  antiprotons, 232–3
  and atomic structure, 97–8, 100, 102, 104–6, 108, 110–13, 118
  sub-structure of quarks, 174–5, 209–12
proximity sensors, 59–60
pulsars, 140, 147
pulse oximeters, 286
Puluj, Ivan, 279–80
Purser, Ken, 96
pyramids, 6, 91

quantisation, 61
quantum biology, 65
quantum chemistry, 65
quantum chromodynamics (QCD), 212

quantum computing, 66, 197
quantum electrodynamics (QED), 58, 145, 207
quantum mechanical tunnelling, 104–5, 122, 128
quantum mechanics, 3, 6–7, 44, 47, 58, 61–3, 65–7, 81–3, 103–4, 112, 122, 164, 173, 193, 235, 291
quarks, 4, 174–5, 180, 200, 208–12, 216–17, 220, 224–7, 249–50, 254, 288, 294, 297, 299
  antiquarks, 211, 288, 297
  colour charge, 212
  discovery of top quark, 233–7, 239, 241, 249
  sea quarks and valence quarks, 211
  tetraquarks and pentaquarks, 262
quasars, 88
quenching, 230, 232

radar, 199–203, 205, 212, 296
radiation-hard detectors, 257
radio astronomy, 139–40, 147, 154
radio emissions, from space, 139–40, 146–7
radio frequency (RF) power sources, 204–5, 207, 214
radio transmitters, 280
radio waves, 52, 139–40, 154, 196, 200–1, 237
radioactive decay, 33, 41–2, 90, 98, 103–4, 114, 181, 189, 194
radioactivity, discovery of, 30–6, 41–4, 49, 67, 120–1
radiocarbon dating, 89, 95, 114–15, 136
radiographs, 15
'radiometric dating', 34
radio-phosphorus, 131–3
radio-sodium, 131–3
radiotherapy, 131–4, 176, 213–16, 238, 257, 298
radium, 31–2, 35, 38, 73, 98–101, 123
Radium Institute, Vienna, 86

rainbows, 47
Ramakrishnan, Venki, 149
Rand, Ayn, 157
Randall, John, 201
Rao, Zihe, 150
Rayleigh, Lord, 35, 53
Raytheon, 202–3
Reber, Grote, 139
rectifiers, 106–7, 112
refraction, 46
Reines, Fred, 183–90, 295
Relativistic Heavy Ion Collider, 244
resonance particles, 173–4, 233
resonators, 284–5
rhumbatron, 200
Richter, Burton, 299
Righi, Augusto, 48
Röntgen, Anna Bertha, 13–14
Röntgen, Wilhelm, 11–16, 19, 25, 27, 47, 272, 280
'Röntgenology', 280
Rossiter, Margaret, 88
Royal Institution, 19–20, 34, 280
Royal Society, 101, 111
Rubin, Meyer, 114–15
Rutherford, Ernest, 29–44, 61–2, 77–8, 129, 141, 143, 273, 291
    and atomic structure, 96–105, 107, 109–12, 118, 121
Rutherford Appleton Laboratory, 113, 229, 255
Rutherford Backscattering Spectrometry (RBS), 116

Samios, Nicholas, 174
SARS-COV-2, 149–50, 152
Saturn, 2, 45
Savannah River Neutrino Experiment, 187
'scanning girls', 173, 175, 293–4
Schuster, Arthur, 287
Schwartz, Melvin, 193
Schwinger, Julian, 145

Schwinger radiation, 146
Schwitters, Roy, 233, 240
Science Museum, 20, 23
Seaborg, Glenn, 133–5, 137
seaborgium, 135
sealing wax, 107
seasonal flu, 152
security scanning systems, 215–16
Segrè, Emilio, 134–5
Segrè chart, 136
semiconductors, 59, 65, 116, 204
SESAME centre, 257
Shane, Donald, 124
Shanghai Synchrotron Radiation Facility, 150
Siemens, Werner von, 13
Siemens-Reiniger-Veifa GmbH, 281
sigma particles, 168, 174
silicon, 65–6, 116
Silicon Valley, 204, 212, 261
skin cancer, 152
Skobeltzyn, Dmitri, 79
Sloane, David, 133
smartphones, 5–6, 60, 116, 153
Smith, Nigel, 196
smoke detectors, 42
Snow, C. P., 41
Snowy Hydro scheme, 90
Soddy, Frederick, 29, 31–6, 42, 281
solar eclipses, 74, 89
solar wind, 89
Solvay Conferences, 129–30
space programmes, 166
SPEAR electron-positron collider, 216–17
speed of light, 164, 209
Spencer, Percy, 203
Sperry Gyroscope, 203
spin, 174
spontaneous symmetry breaking, 250
Standard Model, 3–4, 197, 233–4, 239, 241, 244, 249–51, 262–3, 267, 269–70, 272, 279

Stanford Linear Accelerator Center (SLAC), 208–9, 216, 220, 224, 299
'star noise', 139
'stars of disintegration', 86
Stawell Underground Physics Laboratory (SUPL), 270
Steinberger, Jack, 193
STELLA collaboration, 215, 257
stellar nucleosynthesis, 291
stents, insertion of, 25
Stoletov, Aleksandr, 48
Stone, Robert, 177
strange particles, 163–9, 171, 173–5, 206–7, 297
string theory, 239
strong nuclear force, 4, 85, 168, 180, 193, 195, 206–7, 211–12, 220, 250, 262, 297
structural biology, 148–50
Stuart, David, 148
submarines, 196, 199
Sudbury Neutrino Observatory (SNO/SNOLAB), 191–3, 196–7
Sun, the, 52, 74, 89–90, 140, 196, 249
solar neutrinos, 189–92, 194, 196
Super Proton Synchrotron, 227, 231, 243
superconducting magnets, 228–31, 237–40, 248
Superconducting Super Collider (SSC), 240–1
superheavy elements, 136
Super-Kamiokande, 191–2, 196, 296
supernovae, 88, 140, 147, 194
supersymmetry, 239, 268–9
Sustainable Development Goals, 257
synchrotron radiation, 146–50, 154, 195, 207, 268, 293

Tanaka, Hiroyuki, 91
tau leptons, 192–3, 217, 224–5, 249
tau neutrinos, 233, 241, 249
Taurus (constellation), 147
Taylor, Richard, 209–12, 298

technetium, 120, 134–5, 291
technicolour, 239
Tevatron, 228–9, 231–43, 251, 254–5, 273
theory of everything, 2, 242
thermionic emission, 22
Thomson, Joseph John 'J. J.', 16–23, 25, 27, 29–32, 35–7, 48, 50, 77, 89, 96, 272, 285
thorium, 31–3, 42, 194
Time magazine, 190
Ting, Samuel, 299
Tobias, Cornelius, 137, 177
Tollestrup, Alvin, 229–30, 233
tomography (the word), 281
Tomonaga, Shinichiro, 145
Tonomura, Akira, 64
Torricelli, Evangelista, 26
Treaty of San Francisco, 179
Trezona Formation, 43
triodes, 22, 281
tsunami (2010), 92
Turin Shroud, 43, 114–15
Tuve, Merle, 103, 121–2, 125
type 2 diabetes, 152

Ulam, Stanislaw, 235–6
'ultraviolet catastrophe', 53
United Nations, 257
ununoctium (oganesson), 136
upsilon particle, 225–7, 228, 255
uranium, 31, 34, 42, 103–4, 117, 119, 135, 159–60, 194
Urey, Harold, 128
US Department of Energy, 208, 232, 240
US Geological Survey, 114
US National Bureau of Standards, 147–8

'V' particles, 163
vacuum pump, invention of, 26
Van de Graaff, Robert, 103, 112–13, 298
Varian, Russell and Sigurd, 200–1, 203–4, 209, 212, 298
Veksler, Vladimir, 143–4

Verne, Jules, 15
viruses, 63, 149–52
vitamin B12, 149
volcanoes, 91–2, 194

W and Z bosons, 241, 250, 299
Walker, Sir John, 149
Walton, Ernest, 100–3, 105–6, 108–12,
    116–20, 122, 127–9, 131, 141,
    205, 289
WATCHMAN experiment, 195
water waves, 52
Watson-Watt, Robert, 199
wavelength, human, 63, 286
wave–particle duality, 61–4, 67, 122
weak nuclear force, 4, 168, 180, 193, 220,
    241, 250, 299
weather forecasting, 237
Wembacher, Hertha, 86
Westinghouse, 113, 202–3
whole-body radiation counter, 295
Widerøe, Rolf, 123–4, 205, 291, 299
wigglers and undulators, 151
Wilson, Charles 'C. T. R.', 75–9, 83, 106,
    122, 272
Wilson, Robert Rathbun ('Bob'), 175–7,
    219–24, 227, 230–1, 239, 273, 294
wine fraud, 117

World Wide Web, 5–6, 252–3, 256–7,
    259, 272
Wulf, Theodor, 74
Würzburg Physical Medical Society, 14

X-band accelerating technology, 301
xenon, 137
xi particles, 163, 168, 174
X-ray astronomy, 284
X-ray crystallography, 148–50, 152
X-rays, 52, 86, 146, 148
    and cloud chambers, 77–8
    compared with muons, 90–1
    discovery of, 13–16, 20, 23–6, 47, 49,
    67, 122, 280–1
    high-energy, 53, 106, 133, 141–2, 216
    and radiotherapy, 131–4, 176, 213–14,
    216, 298

Y*(1385) particle, 173
Yanf, Haitao, 150
Yoh, John, 226
Young, Thomas, 45–6, 57
Yukawa, Hideki, 85, 168

zika, 152
Zweig, George, 208, 294

A NOTE ON THE FAIR

A NOTE ON THE TYPE

The text of this book is set in Minion, a digital typeface designed by Robert Slimbach in 1990 for Adobe Systems. The name comes from the traditional naming system for type sizes, in which minion is between nonpareil and brevier. It is inspired by late Renaissance-era type.